Warfare in the 19th Century

HISTORY OF WARFARE

Ian Westwell

A Steck-Vaughn Company

Austin, Texas

www.steck-vaughn.com

Steck-Vaughn Company

First published 1999 by Raintree Steck-Vaughn Publishers,
an imprint of Steck-Vaughn Company.

Library of Congress Cataloging-in-Publication Data

Westwell, Ian
Warfare in the 19th century / Ian Westwell.
p. cm. — (History of warfare)
Includes bibliographical references and index.
Summary: Surveys the changing nature of warfare in the latter half of the nineteenth century, using accounts of various conflicts to describe advances in communication and transportation, changes in battlefield tactics and improvements in weaponry.
ISBN 0-8172-5449-8
1. Military history, Modern--19th century--Juvenile literature.
2. Military art and science--History--19th century--Juvenile literature [1. Military history, Modern--19th century.
2. Military art and science--History--19th century.] I. Title.
II. Series: History of warfare (Austin, Tex.)
D361.W47 1999
355'.009'034--dc21 98-38972
CIP
AC

Printed and bound in the United States
1 2 3 4 5 6 7 8 9 0 IP 03 02 01 00 99 98

Brown Partworks Limited
Managing Editor: Ian Westwell
Senior Designer: Paul Griffin
Picture Researcher: Wendy Verren
Editorial Assistant: Antony Shaw
Cartographer: William le Bihan
Index: Pat Coward

Raintree Steck-Vaughn
Publishing Director: Walter Kossmann
Project Manager: Joyce Spicer
Editor: Shirley Shalit

Front cover: The Battle of Beecher's Island, Colorad September 1868 (main picture) and a Prussian infantryman from the early 1870s (inset).
Page 1: Members of a Boer commando unit in 188 in the Drakensberg Mountains of southern Africa.

Consultant
Dr. Niall Barr, Senior Lecturer,
Royal Military Academy Sandhurst,
Camberley, Surrey, England

Acknowledgments listed on page 80 constitute part of this copyright page.

Contents

Introduction

Warfare underwent very remarkable changes during the second half of the 19th century. In 1840, armies used weapons and fought in ways that would have been familiar to military forces a century before. By 1900, however, warfare had been transformed almost beyond recognition. New weapons, the creation of professional general staffs to plan and conduct wars, and the use of railroads and telegraph systems made wars more planned, controlled, and lethal. Massive firepower and speedy movement became the keys to victory.

Land warfare was transformed by developments in firepower. The muzzle-loading musket gave way to the breech-loading rifle, which could be fired more quickly and was more accurate. The single-shot breechloaders were in turn replaced by rifles that could take magazines (boxes) containing several bullets. The accuracy, range, and volume of fire of these new firearms were all increased. Extra firepower was also being provided by early machine guns.

Artillery was undergoing changes, from muzzle-loading types to breechloaders. Artillery fire became more accurate, and the cannonballs of old gave way to explosive shells. Fitted with fuses, these could be detonated at a set time after leaving the muzzle, or on impact against a target.

Battlefield tactics had to change due to this new firepower or casualties would have been huge. Soldiers no longer fought in close-packed ranks, which made for an easy target. Soldiers began to spread out to reduce the chances of being hit. Frontal attacks were avoided because of the likelihood of very heavy casualties. Uniforms became much darker so that soldiers would blend in with their surroundings. Most major countries still kept large cavalry forces. Cavalry units still made charges, but these often resulted in large numbers of dead and wounded men and horses. Horses were still important for pulling equipment, and would remain so until the development of the internal combustion engine and mechanized transportation.

Major changes also took place in naval warfare. Sail-powered wooden ships were replaced from the 1850s onward with armor-plated, steam-driven warships. The old broadside (row above row of cannon) batteries were replaced by revolving turrets, which were circular armored enclosures carrying two or three guns. Naval gunnery was transformed by explosive shells, which could penetrate armor. New weapons were being developed, including the submarine and underwater torpedo.

The second half of the 19th century saw the world's leading powers, those of Europe, Japan, and the United States, extend their influence through the process of taking over other peoples' lands. Backed by all the resources that industrialized nations could draw on, their campaigns of conquest across the globe were almost always successful. The colonizers suffered the occasional defeats on the battlefield but usually won their wars.

The New Science of War

By the middle of the 19th century it had become clear to a few farsighted senior military figures that the nature of warfare on land and at sea was changing quickly and in many ways. There were many reasons for this, but these changes were chiefly brought about by industrialization and technological progress. These two factors added several new dimensions to warfare. On land, for example, more deadly weapons, especially artillery, and the use of railroads and the telegraph were transforming the nature of conflict.

By the second half of the 19th century countries with more developed economies, and growing populations, could afford to train and equip huge armies, often of hundreds of thousands of men. However, while the soldiers and weapons to fight wars were available, few had considered the ways that conflicts should be conducted or had trained career soldiers to wage them.

One of the first—and the most successful—attempts to adopt a methodical and scientific approach to war planning was by a Prussian, Karl von Clausewitz, a veteran of the Napoleonic Wars. His greatest book was *On War*, begun in 1819. He analyzed his own experiences of war and attempted to create a methodical approach to warfare. Clausewitz's approach was very influential, particularly in his homeland where one man, Field Marshal Helmuth von Moltke, developed his ideas to a greater extent.

Helmuth von Moltke was the man who founded the Prussian General Staff, which was responsible for planning and running wars.

Detailed planning

Moltke, a Prussian aristocrat, set about completely overhauling Prussia's military structure He reorganized the Prussian General Staff (the organization responsible for running the country's military affairs) into four separate units: three military districts

Prussian artillerymen take part in peacetime maneuvers. By training regularly, the Prussians made sure that they were well prepared for action when war broke out.

(West, East, and German) and the Railroads Department. His aim was to make the state's armed forces more efficient and effective. Prussia was an aggressive state and looked to extend its control over central Europe. War was the chosen method. Moltke decided that the best way to defeat an enemy was to plan, down to the smallest detail, the strategy to crush an enemy before war was even declared.

Using large maps and wooden blocks to represent Prussia's and an enemy's armies, Moltke and members of the General Staff would conduct "war games," trying to model the best way to win a victory. Once a general plan of campaign had been agreed upon, the necessary practical arrangements could be made. These would include such factors as where armies should be massed for best impact, how and when they needed to move and attack, and what they had to achieve by a certain time. This was, in reality, war by timetable. Every army committed to a war was expected to achieve a particular objective—capture a city, or destroy an enemy force, for example—according to the prearranged plan.

Using railroads Moltke was able to move his forces quickly and thanks to the telegraph he was kept informed of their progress and any difficulties that might arise. Individual army commanders, although they had a degree of independence to make their own decisions, could always ask for help from Moltke and the General Staff. Moltke's ideas were first tried out in Prussia's 1866 war with Austria and later proved stunningly successful during the Franco–Prussian War (1870–71).

A nation ready for war

During Moltke's time Prussia also developed one of the world's mightiest war machines. Soldiers usually received the most up-to-date weapons, their officers underwent professional training in their duties, and the state devoted much of its resources to improving its war-fighting ability. All males of military age, for example, had to undergo military training for a set period, and then had to practice their military skills for a few weeks each year.

In times of war these men, known as reservists, could quickly rejoin their units, be given weapons and equipment, and sent off to fight—in a matter of weeks, if not days. This process ensured that the Prussians had a head start on their military rivals, none of whom had such efficient plans for mobilization.

All of the resources of the Prussian state—human, economic, scientific, and technical—could be used to fight an enemy. Moltke and the Prussian General Staff established a systematic approach to conflict that remains the foundation of military thought and planning for large-scale war in the modern world.

ALFRED THAYER MAHAN

Field Marshal Helmuth von Moltke and the Prussian General Staff were chiefly concerned with fighting wars on land, but the American military thinker Alfred Thayer Mahan developed similarly thoughtful ideas on the use of seapower. His key book, published in 1890, was *The Influence of Seapower Upon History, 1660–1783*. This was a study of naval strategy and was used as the basis for discussing the role of navies in the modern world.

Mahan argued that seapower was vital to national security and that for a nation to keep its security in wartime it was essential to smash an enemy's fleet. However, Mahan asked broader questions about the link between a country's political goals and naval strength. As he famously remarked in 1892: "All the world knows, gentlemen, that we are building a new navy. Well, when we get our new navy, what are we going to do with it?"

Mahan was stating that it was for politicians to decide what a navy's strength and objectives should be but that naval officers should be free to decide how to achieve those objectives.

THE CRIMEAN WAR

The Crimean War was fought from 1853 to 1856 by a number of countries led by Britain and France against the Russians. It was probably the most badly managed war of the 19th century. The commanders of both sides were of poor quality, the Russian forces were badly trained, and the British and French forces suffered heavy casualties from disease and hunger. The war is perhaps best remembered for a military catastrophe—the valiant but suicidal cavalry attack known as the "Charge of the Light Brigade."

The Crimean War was really about the balance of power in Europe but it began over who had control of Jerusalem. Turkey controlled the city but Russia was eager to extend its influence in the region. France sided with the Turks when the Russians took over Wallachia and Moldavia (now parts of Romania, but then part of the Turkish Ottoman Empire). Britain supported France, and both sent fleets to defend Constantinople, the Turkish capital. The Russians controlled the Black Sea from their base at Sevastopol in the Crimea and wanted to take over the Turkish Dardanelles, the only sea route from the Black Sea to the Mediterranean. Turkey declared war on Russia in October 1853.

The opening moves

The Turks attempted to push the Russians out of Moldavia and Wallachia, winning the Battle of Oltenitza on November 4, 1853. But this was followed by a Russian naval victory a few weeks later.

Russian warships attack the Turkish fleet at anchor in the harbor of the Black Sea port of Sinop, November 30, 1853. After just two hours all of the Turkish vessels were sunk.

THE BATTLE OF SINOP

The battle between the Russians and Turks at the Turkish port of Sinop in the southern Black Sea on November 30, 1853, was one of the few major sea battles of the 19th century. The Turkish fleet, ten warships commanded by Osman Pasha, faced a Russian squadron of eight vessels under Admiral Pavel Nakhimov.

Nakhimov, however, had a decisive advantage. Many of his warships, although sail-powered like those of the Turks, had cannon that fired explosive shells, rather than cannonballs. These could penetrate the sides of wooden vessels and then explode with devastating impact, killing crewmen and destroying cannon. Both fleets consisted of wooden ships but Nakhimov's cannon were more effective, setting fire to many of the Turkish ships.

The battle began at noon and the Turkish fleet was totally destroyed within two hours. Sinop was then attacked and surrendered to the Russians four hours later. The immediate outcome of the one-sided battle was that Britain and France allied with Turkey to prevent any further Russian attempts to dominate the Black Sea, thereby sparking the Crimean War.

The Battle of Sinop on November 30 saw the Russian fleet smash the Turks. The Russians next invaded Bulgaria, then ruled by the Ottomans, and laid siege to Silistria. Both Britain and France declared war on Russia in March 1854. They sent a joint army to Varna in Bulgaria. Other European countries felt threatened by Russia's invasion of Bulgaria. The Austrians sent troops into the Danube River area. The Russians abandoned their siege of Silistria and withdrew completely from Bulgaria in early August.

The Russians rejected the allied peace proposals offered after Silistria. The French and British then decided to destroy Russia's power in the Black Sea. The port of Sevastopol in the Crimean peninsula had to be captured. Neither the French nor the British were prepared to fight so far from home. Their key problem would be keeping their men supplied and protecting them from the intense winter cold. To make matters worse, no single general was in charge of the French and British forces. The expedition to the Crimea was commanded by the British Lord Fitzroy Raglan, a 66-year-old veteran, and the French Armand de Saint-Arnaud, who was suffering from a very serious illness.

The French and British left Varna in a convoy of over 150 warships on September 7 and landed on the Crimean peninsula at Calamita Bay during the middle of the month. Sevastopol lay

30 miles (48 km) from the landing beach and the allies—50,000 British, French, and Turkish infantrymen, 128 cannon, and 1,000 British cavalrymen—began their advance on September 19. The Russian commander, Prince Alexander Menshikov, marched his 37,000 troops to oppose the allies.

Finding secure bases

The two armies met along the Alma River on September 20. Menshikov placed his troops along the riverbank and on the high ground beyond. The allies were able to cross the river without much difficulty but the British infantry had to fight hard to push the Russians off the high ground. The allies suffered 3,000 casualties, while the Russians had 6,000. The road to Sevastopol was open. Before the allies could lay siege to the port, however, they had to capture a sheltered port themselves. They needed a secure base to bring their heavy artillery ashore, to provide shelter for their troops, and a safe anchorage from the worst of the winter storms. Two anchorages were chosen, one at Kamiesch for the French and the other at Balaklava for the British.

Kamiesch and Balaklava were to the south of Sevastopol. The allies had to march around the port, abandoning their line of retreat to their original landing beaches. It was a tricky maneuver but the Russians did not attack. Menshikov, who had withdrawn into Sevastopol after the Alma battle, led his army north from the port to meet Russian reinforcements heading for the peninsula.

The defenses of Sevastopol, particularly those facing Kamiesch and Balaklava, were far from complete when the siege opened in early October. The allies failed to take advantage of this and

Lines of British troops advance against the Russian army at a key moment in the Battle of the Alma River on September 20, 1854. The Russians were finally forced to retreat after putting up a stiff fight.

The Crimean War was fought between 1853 and 1856. The key issue was over who should control the Black Sea, Russia or Turkey. Most of the war was fought around the Russian port of Sevastopol.

allowed the commander of the engineers building the Russian defenses, Colonel Franz Todleben, time to strengthen his position. The allied siege was delayed at this vital time because of the death of Saint-Arnaud on September 29 and their lack of the necessary engineers and equipment to begin such an operation. Saint-Arnaud's replacement was General François Canrobert.

The Charge of the Light Brigade

The allied bombardment of Sevastopol began on October 17, but the allies faced a more pressing task—Menshikov was attempting to drive a wedge between the allied troops besieging Sevastopol and their main base at Balaklava. The Battle of Balaklava on October 25 was, in fact, several battles in one. The Russians were able to capture a number of hilltop fortifications held by the Turks but their follow-up cavalry attacks were repulsed by a brilliant charge by the heavily outnumbered British Heavy Brigade (cavalry) and the volley fire of the 93rd Highlanders, the famous "Thin Red Line" (red was the color of the Highlanders' jackets).

However, the next phase of the battle was a disaster for the allies. Possibly due to a poorly worded written order, the foolishness of a young officer who identified the wrong Russian position to attack, or the pigheadedness of their commander, Lord James Cardigan, the British Light Brigade, charged the main Russian army. The Russians were arrayed on three sides of the valley down which the brigade charged. As the British cavalrymen advanced, they were first blasted by cannon and then at closer range by rifle fire. Remarkably the British cavalrymen reached the Russian artillery a mile away from their start position, but they were then counterattacked by fresh Russian cavalry.

The survivors, many wounded, had to retreat. Losses were heavy—673 officers and men had begun the charge; 247 men never returned. Nearly 500 horses were lost. Cardigan himself escaped without a scratch. Only a superb charge by units of French cavalry prevented losses from being even greater during the Light Brigade's retreat. At a stroke the allies had lost probably their finest cavalry unit. As one French general remarked of the charge: "It is magnificent, but it is not war."

The Battle of Balaklava was a draw. The allies held on to Balaklava, but the Russians had captured high ground to the north of the port. Menshikov decided to try for Balaklava again on November 5. The Battle of Inkerman was badly directed by the rival generals. The fighting, mainly between the British and

Cavalrymen of the British Light Brigade launch their disastrous charge during the Battle of Balaklava on October 25, 1854. Blasted by Russian cannon and rifles, the brigade was cut to pieces in just 20 minutes. About a third of the British cavalrymen were killed in the attack.

FLORENCE NIGHTINGALE

Almost singlehandedly Florence Nightingale revolutionized the medical care of soldiers suffering from battle wounds, illness, and disease. Nightingale trained as a nurse in the early 1850s, and was appalled by newspaper reports of the horrible conditions that the British forces committed to the Crimean War faced. Nightingale volunteered to tend the troops but was given charge of the military hospital at Scutari, near Constantinople.

Nightingale arrived at Scutari on November 5, 1854. She immediately set about cleaning the hospital and making conditions more hygienic. Floors and walls were scrubbed, clean bedding was made available, and soldiers were given proper care and attention by Nightingale and her band of nurses.

Recovery rates in the hospital improved dramatically as deaths from infections brought on by the previously poor medical care of wounds dropped. Nightingale also toured Scutari late at night, offering sympathetic words to the sick and wounded. She became known as "The Lady with the Lamp."

In March 1856, Nightingale was made the general superintendent of the Female Nursing Establishment of the Military Hospitals of the Army, and used her influence to improve conditions for the army's ordinary soldiers. Acting on her recommendations, the Army Medical School was founded in 1857. Henceforth, soldiers knew that they stood a much better chance of surviving the horrors of war. The high quality of medical care demanded by Nightingale led to these changes.

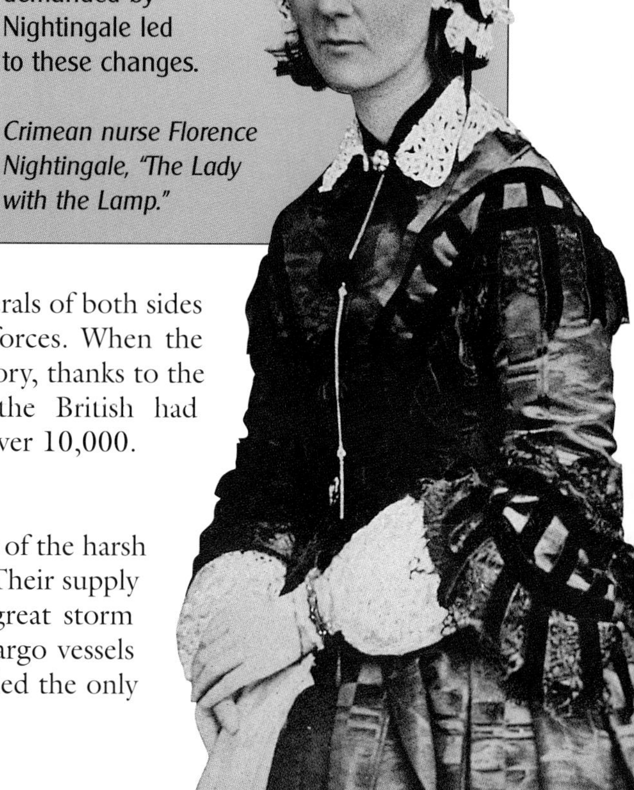

Crimean nurse Florence Nightingale, "The Lady with the Lamp."

Russians, lasted most of the day and the generals of both sides failed to exercise much control over their forces. When the slogging match ended in a narrow allied victory, thanks to the timely arrival of French reinforcements, the British had around 3,000 casualties. The Russians had over 10,000.

Disease, cold, and hunger

Both sides settled down to ride out the worst of the harsh winter. It was a terrible time for the British. Their supply system was already badly organized and a great storm added to their problems by destroying 30 cargo vessels on November 14. The Russians also controlled the only

British siege artillery sited on high ground overlooking the walls of Sevastopol fires on the Russian-held port. Once the port's defenses had been smashed, British and French troops attacked in strength.

decent road from Balaklava to the siege lines around Sevastopol so the men there suffered from severe food shortages as the allies had to drag supplies over difficult muddy ground. Most of the troops lacked shelter or winter clothing. The medical facilities were limited and disease, hunger, and cold killed many soldiers. The allies did receive some much-needed reinforcements—10,000 Sardinian troops—in mid-February 1855. Due to public anger back in Britain the supply system gradually improved and thanks to the outstanding work of Florence Nightingale and her nurses in the main hospital at Scutari, near Constantinople, medical facilities improved.

The first battle of the new year took place on February 17. The new Russian commander in the Crimea, Prince Michael Gorchakov, launched his field army against a road and railroad the allies were building from Balaklava to their troops outside Sevastopol. The Battle of Eupatoria was a halfhearted affair and a Turkish force stopped the Russian attack in its tracks. The allies took heart from this and intensified their siege operations. A massive bombardment between April 8 and 18 smashed much of Sevastopol's defenses. The Russians expected a full-blooded allied assault, but it never came. The allied commanders—and their governments—were arguing over how best to fight the war. Canrobert, furious at the squabbling, resigned. He was replaced by General Aimable Pélissier.

The battle for Sevastopol

The allies did gain a major advantage on May 24, when a successful operation against Kerch on the eastern seaboard of the peninsula ended the flow of Russian supplies to the Crimea across the Sea of Azov. Around Sevastopol itself the allies finally solved their difficulties and launched a successful attack on the port's outer defenses on June 7.

They next turned their attention to attacking two of the key forts protecting Sevastopol—the French attacked the Malakoff and the British assaulted the Redan. The assaults on the night of June 17–18 were a disaster. The French troops launched an uncoordinated assault and were shot to pieces by Russian rifle fire; the British columns were smashed in a deadly crossfire from over 100 cannon. The allies suffered 4,000 casualties; the Russians slightly fewer. The British commander, Lord Raglan, died a few days later. He was replaced by General James Simpson.

After the complete failure of the night attack the allies settled down to bombard Sevastopol into submission. The Russian garrison's losses reached over 300 troops a day. The port would fall if a relief force could not break through. Gorchakov launched part of his field army against 35,000 allied troops holding high ground overlooking a small river on August 16. The fight, known as the Battle of the Traktir Bridge, lasted five hours. Gorchakov retreated and the noose tightened around Sevastopol.

Sevastopol falls

The allies planned a second attack on the Malakoff, September 8. This was preceded by a massive three-day bombardment. Then, at noon on the 8th, the French swarmed out of their trenches less than 30 yards (29m) from the Malakoff, and captured the fort after some stiff hand-to-hand combat. The British attack on the Redan was halted by the Russians. The French in the Malakoff turned its guns on the Redan and the Russian troops there retreated. Without the Malakoff and the Redan, Sevastopol could not survive. Gorchakov ordered the port's garrison evacuated and its fortifications destroyed.

The war was effectively ended by the fall of Sevastopol but dragged on in minor operations until February 1856. A peace treaty was agreed upon in March. The war had produced enormous casualties. Both sides had about 250,000 casualties in the war but only 70,000 of the allied casualties were due to wounds. The rest died from disease, hunger, and cold. Half of the Russian casualties died from similar causes.

The Crimea did, however, mark a turning point in this respect. Henceforth, European armies paid close attention to the wellbeing of their soldiers. The provision of food, clothing, shelter, and medical aid became central to war planning.

French officers encourage their troops to continue their attack on the Russian defenders of the Malakoff fort. The fort was a key point of Sevastopol's defenses. Its capture on September 8, 1855, forced the Russians to abandon the port.

THE INDIAN MUTINY

By the mid-19th century the British controlled India, but their rule of the subcontinent was not as secure as they thought. In the past the British had used rivalries among the local kings and princes they allowed to rule to prevent the Indians from uniting against them. The British had also built up an army containing over seven times as many seemingly loyal Indian as European troops. However, in 1857, an incident at an Indian army garrison sparked an uprising—the Indian Mutiny—that almost ended British rule.

The incident at the garrison of Meerut on May 10, 1857, had been simmering for some time. The British officers who led Indian regiments often treated their sepoys (local Indian soldiers) badly. To make matters worse there was a rumor that the British had introduced a new paper rifle cartridge covered in either cow or pig fat to protect it from moisture. Before the cartridge could be used it had to be bitten open. However, cows are sacred to Hindus and pigs are considered "unclean" by Muslims.

These soldiers are dressed in the colorful uniforms worn on parade by sepoys serving with the British at the time of the 1857 mutiny.

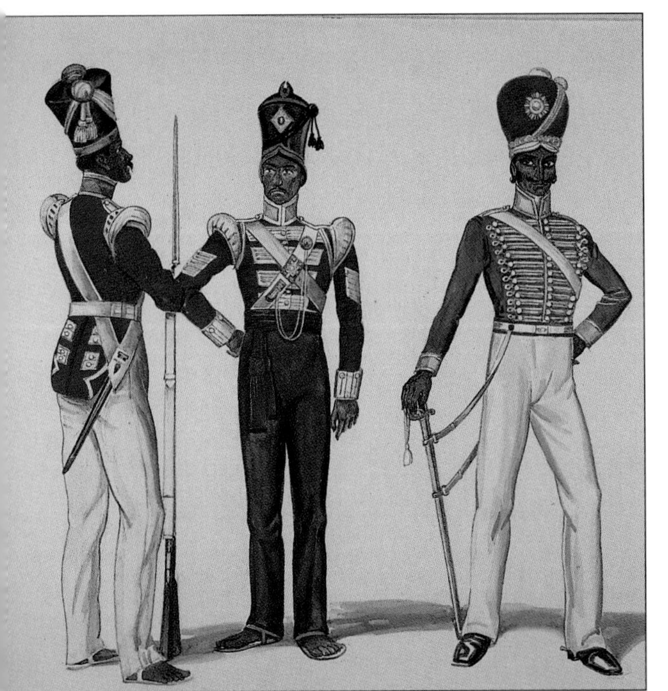

Widespread atrocities

More than 80 of the Meerut sepoys refused to bite into the new cartridges and were imprisoned for disobeying orders. They were rescued by their fellow sepoys, who then attacked their British officers and wives while they were at Sunday prayer. What followed was the first of many atrocities committed by both sides. Most of the Europeans were killed and the sepoys then headed for Delhi, some 25 miles (40 km) away.

The Indian troops in Delhi and many local civilians joined the Meerut sepoys, declaring the Mughal ruler Bahadur Shah their leader. The Mughals were Muslims who had carved out an empire in India. Many Europeans in Delhi were slaughtered, but a few found sanctuary in the British compound outside the city, before fleeing to Meerut and Umballa. The news of the uprising spread across India and similar revolts occurred elsewhere.

The British responded by sending a weak force—about 3,000 men—to Delhi but they had little immediate hope of recapturing the city. Elsewhere other British garrisons were besieged by the rebels. At Lucknow Sir Henry Lawrence had just 1,720 men, including 712 loyal sepoys, to protect the town and over 1,200 noncombatants. Sir Hugh Wheeler at Cawnpore had far fewer troops to protect 200 noncombatants.

Massacres at Cawnpore

Cawnpore was besieged by the rebels for three weeks and Wheeler, seeing no hope of relief, agreed to surrender terms in late June. On the 27th, his garrison and the noncombatants, mostly British women and children, marched out with a promise of safe passage to Allahabad. They never reached safety. While the refugees were embarking on boats, the rebels opened fire, killing many of them. The few women survivors and their children were imprisoned but then murdered on July 15. Their bodies were thrown down a well.

There was soon better news for the British. Sir Henry Havelock at the head of 2,500 troops had marched to the relief of Lucknow in early July. In nine days (July 7–16) Havelock traveled 125 miles (200 km) and defeated rebels in three battles. The last battle saw the British retake Cawnpore. Inside the city they found the butchered remains of the European prisoners at the bottom of the well. The British vented their fury, slaughtering many Indians, rebels or not. Havelock waited for reinforcements to arrive at Cawnpore and resumed his march on Lucknow on September 20.

A BRUTAL WAR

The Indian Mutiny was characterized by great brutality. Atrocities against civilians and prisoners were committed by both sides. The mutiny by Indian troops in the British base at Meerut on May 10, 1857, which sparked the war, involved the murder of European civilians, including women and children. Similar scenes of murder and arson also took place when Delhi fell to the mutineers. The worst incident occurred at Cawnpore, where the British garrison and European civilians were massacred in June and July.

However, the British behaved no better. When they recaptured Cawnpore, the British general ordered all of the captured mutineers to each in turn clean up the blood-splattered walls and floors of the house in which the garrison's women and children had been murdered. When each mutineer had completed his allotted task, he was taken outside and hanged.

The British also executed mutineers by tying them directly in front of cannon. Each cannon was fired and the mutineer would be blown apart by the cannonball. This type of execution was deliberately cruel as the mutineers believed that they could not have an afterlife if their bodies were not whole.

The British won a notable victory at Delhi on the same day. The initial force of 3,000 men besieging the city had captured some key high ground on June 8, but had to wait for heavy artillery and reinforcements before attacking Delhi's great walls. Some 4,000 British troops stormed the walls on September 14 but it took a week to clear the city of rebels. Bahadur Shah, the head of the mutiny, was captured and executed. The fall of Delhi was a major blow against the rebels, but the fighting continued.

Ending the rebellion

Attention focused on Lucknow. Havelock managed to cut a path through the 60,000 Indian besiegers but lost 25 percent of his troops on September 25. However, he and the garrison were too weak to defeat the besiegers. The siege dragged on for six more weeks. Many died from hunger, disease, and exhaustion. A British relief force from Cawnpore under Sir Colin Campbell arrived outside Lucknow on November 14, broke through two days later, and had withdrawn back to Cawnpore by the 22nd.

The British at Cawnpore under Campbell spent months waiting for reinforcements before mounting an expedition to drive the rebels out of Lucknow once and for all. The main addition to

British troops storm Delhi's Kashmir Gate one of the main gateways into the city, on September 14, 1857. Casualties among the 4,000 British attackers were heavy and it took until the 20th to drive the mutineers out of the city.

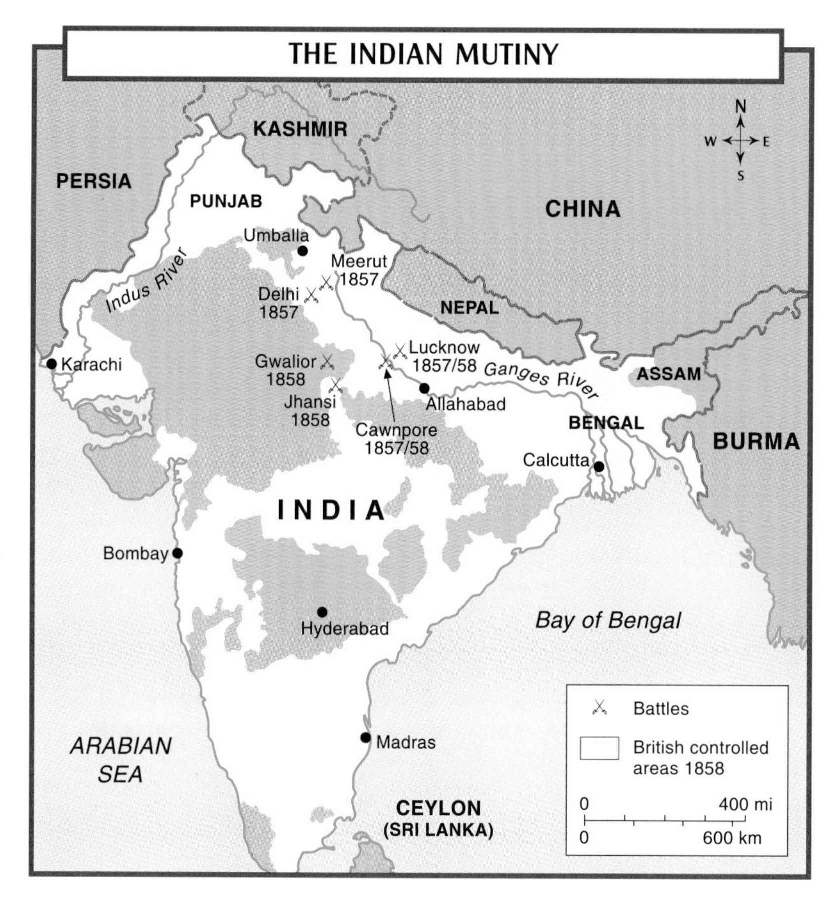

The Indian Mutiny lasted from 1857 to 1858 and came close to ending British rule in India. However, the British were able to defeat the mutineers and recapture several key cities, despite being heavily outnumbered.

Campbell's command were 10,000 Gurkha soldiers from Nepal, who were paid to fight for the British. After a week of bitter fighting Campbell's forces retook Lucknow on March 16, 1858.

The first three months of 1858 saw British forces under Sir Hugh Rose regain control of central India. Rose inflicted two defeats on the mutineers in February and March. He then advanced on the city of Jhansi, a center of rebellion, and placed it under siege. Rose defeated a rebel force of 20,000 on April 1 and captured the city two days later.

The rebels were defeated by Rose on several other occasions, over the following weeks but his decisive victory came at Gwalior on June 19. Of the two rebel leaders present, one, the female Rani of Jhansi, was killed and the other, Tantia Topi, fled. Topi was captured and executed on April 18, 1859, but the Indian Mutiny was already over. British rule in India would continue into the next century.

ITALY'S WARS OF UNIFICATION

At the end of the Napoleonic Wars in 1815 Italy consisted of a number of independent states and kingdoms controlled by monarchs who cared little for their subjects. Italy was also dominated by the Austro-Hungarian Empire, which had huge economic and political influence in the peninsula's affairs. Some Italians, however, wanted to end this interference and replace the independent states with a strong unified country. Resistance to the Austro-Hungarians focused on the Italian state of Piedmont.

Piedmont allied with France, one of the Austro-Hungarian Empire's rivals, in March 1859. The war that followed was inconclusive and did nothing to satisfy the Italian nationalists. It is perhaps best remembered for the Battles of Magenta (June 4) and Solferino (June 24) in northern Italy. Both sides lost heavily. The blood soaking through the blue uniforms of the French troops at Magenta gave rise to the naming of the purplish-red color. One man, Swiss Jean Henri Dunant, was so horrified by the suffering of the wounded at Solferino that he founded the Red Cross.

The "Thousand Redshirts"

On May 11, 1860, a group of Italian nationalists, the "Thousand Redshirts," (they wore red shirts) led by Giuseppe Garibaldi landed on the island of Sicily, part of the Kingdom of Naples. Garibaldi had the backing of Piedmont's king, Emmanuel II, and

The French emperor, Napoleon III (center), watches his troops go into battle against the Austro-Hungarian army at Solferino on June 24, 1859.

Garibaldi's 1,000 Italian nationalists embark on boats at the northern Italian port of Genoa to begin their journey to Sicily, May 6, 1860. Garibaldi planned to capture the island, which was part of the Kingdom of Naples, as the first stage in his campaign to unify the whole of Italy.

the kingdom's premier, Camillo di Cavour. Garibaldi completely defeated the Neapolitan troops on the island. With a secure base Garibaldi crossed to the Italian mainland on August 22.

The Redshirts captured Naples on September 7. Piedmont then invaded the Papal States (lands controlled by the pope centered on Rome). Garibaldi laid siege to the Neapolitan port of Gaeta, which fell on February 13, 1860. The Kingdom of Naples surrendered. A united Italy was proclaimed on March 17, 1861. The Papal States, however, were occupied by the French. Garibaldi tried to make the Papal States part of Italy, but the Italian government did not want to upset the French. Forces loyal to the government defeated Garibaldi in August 1862.

Italy united

Garibaldi made another attempt to bring the Papal States under Italian control in 1866, after the French had withdrawn from Rome. He did not have the backing of the Italian government and failed. Nevertheless, he tried again in 1867. However, the French returned and Garibaldi was defeated again.

Rome did eventually become part of Italy in 1870, but Garibaldi was not involved. The French had to withdraw their garrison, which was needed in its war against Prussia. A 60,000-strong Italian army captured the city on September 20. Less than two weeks later Rome was formerly incorporated into Italy, becoming its capital. The wars of unification were over.

PRUSSIA'S RISE TO POWER

By the mid-19th century Prussia was the most ambitious German state in central Europe. The Holy Roman Empire, once the dominant power in the region, had been dissolved by Napoleon Bonaparte in 1806. Its replacement, the Austro-Hungarian Empire, was far from united due to tensions between its different peoples. Prussia became the main rival to Austria-Hungary. The Prussian politician Count Otto von Bismarck tried to make Prussia the heart of a great German empire. War between the two was almost inevitable.

Before taking on Austria-Hungary, Prussia tried to take over some of the region's less powerful states. Prussia had already tried to take over the Danish province of Schleswig-Holstein in 1848 but backed down in the face of a coalition (temporary alliance) of Austria, Britain, Russia, and Sweden. By the beginning of 1864 Prussia was ready to try again.

Prussia goes to war

The invasion of Denmark began on February 1, 1864. The Prussians swept all before them. A truce was signed on April 17. However, Bismarck succeeded in preventing a peace treaty being signed. He believed Prussia might be forced once again to give up the province. The war resumed and in early August Denmark was forced to surender Schleswig-Holstein to Prussia.

Bismarck had succeeded in making Prussia the dominant power along the northern coast of central Europe and began to plan for the defeat of Austria-Hungary. In 1865 Bismarck provoked a crisis with Austria-Hungary. Austria-Hungary objected to Prussia's continued occupation of Schleswig-Holstein and Prussia's treaty with France, which had recently been at war with Austria-Hungary. Several other German states, which feared Prussia's growing power, supported Austria-Hungary.

In June 1866 Prussia declared war on Austria-Hungary and on those states in Germany—Bavaria, Hanover, and Saxony—that supported Austria-Hungary. Italy, allied to Prussia, also declared war on Austria-Hungary. An Italian army was crushed by the Austro-Hungarians at Custozza on June 24 and an Italian fleet was defeated off the island of Lissa (now Vis) in the central part of the Adriatic Sea on July 20.

THE BATTLE OF LISSA

On July 20, 1866, an Austrian fleet commanded by Admiral Wilhelm von Tegetthoff decisively defeated an Italian force led by Admiral Carlo di Persano close to the island of Lissa (now Vis) in the central Adriatic Sea. Tegetthoff led his seven ironclad warships and 14 wooden vessels straight at Persano's ten ironclad warships and 22 wooden vessels.

Apart from their numerical advantage, the Italians also had the more modern cannon mounted on their vessels. Tegetthoff, reasoning his ships would be blown to pieces in any exchange of long-range fire, decided to get to close range as quickly as possible. His ships' older cannon would then stand a chance of damaging the Italian ironclad warships.

The Italians opened fire on the Austrians as they attacked, but did little damage. Tegetthoff opened fire at point-blank range, and the battle deteriorated into a confused ship-against-ship struggle. The tide turned when Tegetthoff's flagship, the *Ferdinand Max*, rammed and sank the *Re d'Italia*. The Italians fled.

The world's navies drew the wrong conclusions from Lissa, believing that rams still had a part to play in naval warfare. For the next 30 years navies fitted rams to their warships and practiced ramming. In fact Lissa was unique. As naval guns increased in accuracy and power, ramming was a far too dangerous tactic. Ships would be blasted out of the water before they could ram an enemy.

An Italian ship explodes in flames at the height of the Battle of Lissa on July 20, 1866. The recently founded Italian navy was defeated by an Austro-Hungarian fleet. The Italians had two of their warships sunk in the battle.

The war that followed, the Austro-Prussian War, is sometimes known as the Seven Weeks War. It totally justified Bismarck's political maneuverings and his faith in the organizational abilities of General Helmuth von Moltke, head of the Prussian General

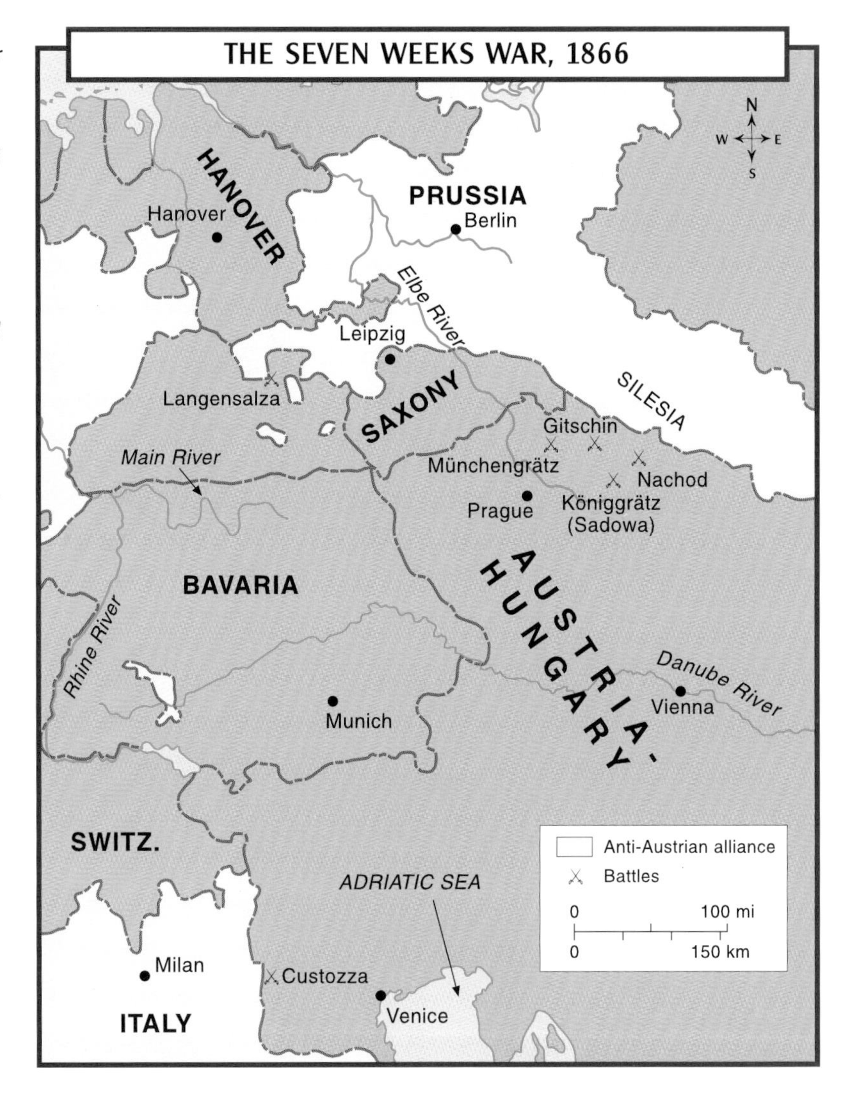

The Seven Weeks War in 1866 was chiefly fought between the armies of Prussia and the Austro-Hungarian Empire. The Prussian army was by far the superior and its generals had planned the war down to the smallest detail, making sure that Austria-Hungary's defeat was both rapid and total.

Staff. Moltke's arrangements were complex but planned down to the finest detail. On July 16, one Prussian force, 50,000 men under General Vogel von Falkenstein, attacked the state of Hanover from the west. Three other armies marched through Silesia and Saxony.

One of the first battles of the war took place in west Germany. A Hanoverian army initially pushed back part of Falkenstein's force on the first day of the Battle of Langensalza on June 27 but was surrounded and forced to surrender two days later. However, this was a diversion; the Prussians had planned that the main campaign of the war would take place in the southeast.

The Prussians outmaneuvered their opponents in the southeast. The Prussian Army of the Elbe and the First Army linked up and cut through advance elements of the main Austro-Hungarian army at the Battles of Münchengrätz on June 27 and Gitschin two days later. Another Prussian army, the Second, won a victory at Nachod on the 27th and then pushed toward Gitschin.

The combined Prussian armies of some 220,000 men now sought to bring the main Austro-Hungarian army to battle. By means of the telegraph Moltke ordered his commanders to carry out a maneuver to trap the Austro-Hungarians. The maneuver did not work, but the battle that followed—Königgrätz—was decisive.

Prussian infantrymen come under heavy fire from the Austrian defenders of the village of Chlum during the decisive Battle of Königgrätz on July 3, 1866.

Prussia unites the German states

The Battle of Königgrätz (also known as Sadowa) on July 3 did not run smoothly for the Prussians. Their attacks were met by ferocious Austro-Hungarian counterattacks. The first Prussian attack was by the Army of the Elbe and the First Army, and Moltke ordered the Second Army to rush to their aid. The Second Army attacked the northern section of the Austro-Hungarian line in the early afternoon. The Prussians then pounded away at the center of the Austro-Hungarian position with their artillery. The Austrians began to withdraw. The Austro-Hungarians had 45,000 casualties, the Prussians 10,000.

Victory at Königgrätz ensured that Prussia would triumph over Austria-Hungary. The Treaty of Prague that was signed on August 23 was wholly favorable to Prussia. Prussia became the head of a new North German Confederation containing all the German states north of the Main River. Austria-Hungary was forbidden from becoming involved in German affairs, and German states south of the Main River were allowed to form the South German Confederation.

Bismarck had made Prussia the dominant state in central Europe partly due to his political skills, which allowed him to exploit the rivalries between other European powers to his own advantage. When diplomacy gave way to war, Prussia had an army that was superior to any of its rivals, as the war of 1866 proved.

THE FRANCO–PRUSSIAN WAR

France viewed the growing might of Prussia in the 1850s and 1860s with increasing concern. Due to Count Otto von Bismarck's political skill and the might of the state's armed forces, Prussia had become the dominant power in central Europe. Bismarck, however, wanted more. He decided that the best way to create a united Germany led by Prussia was to go to war with Prussia's chief rival in Europe—France. By early 1870 Bismarck had engineered a political crisis that made war between France and Prussia certain.

In fact it was France that declared war on Prussia. The French emperor, Napoleon III, and the French government became alarmed in the middle of 1870 when the Prussians tried to put a Hohenzollern on the Spanish throne. Hohenzollern was the name of the Prussian royal family. The French believed that with Hohenzollerns on both the Spanish and Prussian thrones they would face a war on two fronts at some time in the near future. On July 15, 1870, France declared war on Prussia. A number of other German states—Baden, Bavaria, and Württemberg—sided with Prussia a day later.

Superb mobilization plans

The Prussian General Staff under the direction of General Helmuth von Moltke had been planning for war against France for some time. The Prussian mobilization plans, which involved the movement of hundreds of thousands of troops by railroad to the border with France, ran like clockwork. The French did not have such

A Prussian infantryman takes aim with his breech-loading Dreyse rifle during a skirmish with the French in 1870.

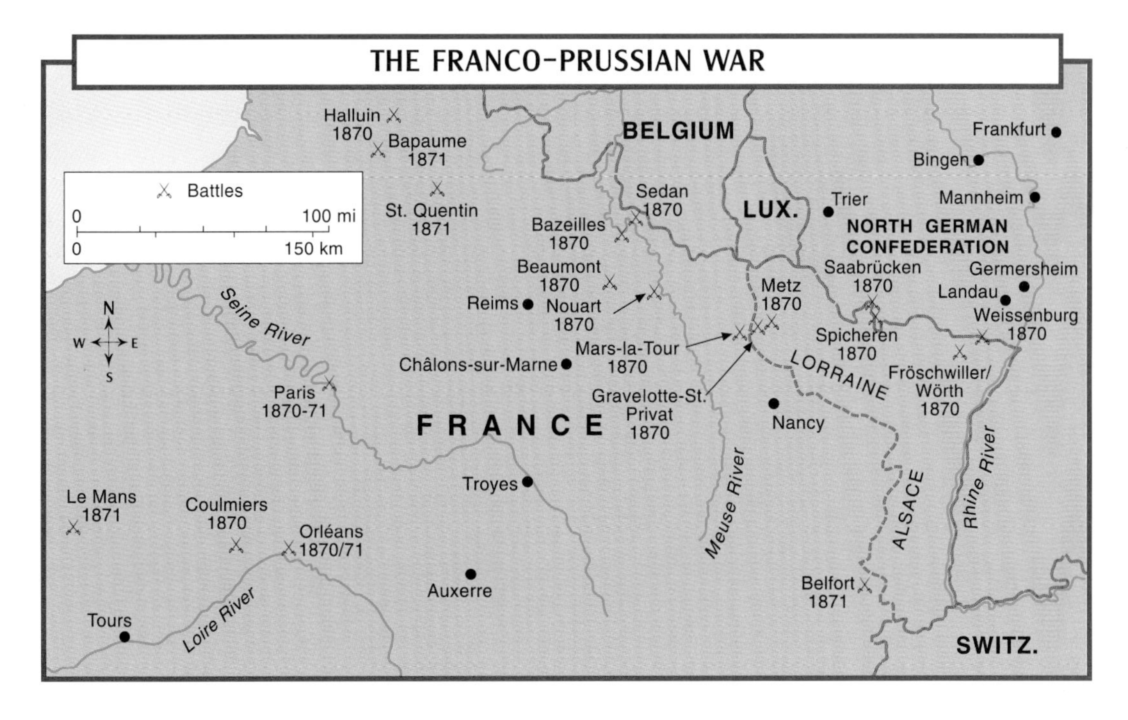

well-oiled plans and their mobilization was haphazard, slow, and incomplete. The Prussians had also been able to find out the complete order of battle for all of the French armed forces.

The 1870 war between France and Prussia fell into three stages. First, the Prussians defeated the main French field armies in a series of battles in eastern France. Second, the Prussians besieged Paris. Third, the Prussians had to deal with newly raised French armies.

The Prussians had a strategy to win the war quickly. Three armies totaling close to 400,000 men were on the French border by the end of July. One army, 60,000 men, was between Trier and Saarbrücken; 175,000 troops were placed around Bingen and Mannheim; and 145,000 troops were between Landau and Germersheim. Before the fighting had begun, the Prussians had overwhelming forces at key points on the border.

The opening battles

Because of their slow mobilization, only 220,000 French troops were massing against the Prussians. These were divided among eight small armies along the frontier and at fortresses near the border. The French armies were in danger of being defeated one after the other. Napoleon and his war minister, Marshal Edmond Leboeuf, were in command, but they were not in the same class as Moltke and his generals.

The first battle of the war took place at Saarbrücken on August 2, 1870. It was a minor affair but led to the reorganization of the French. The eight small armies were consolidated into two bigger commands: the Army of Alsace led by Marshal Marie MacMahon and the Army of Lorraine under Marshal Achille Bazaine. While these changes were taking place, the Prussian Third Army surprised part of MacMahon's forces. The French were outnumbered at the Battle of Weissenburg on August 4 and had more than 1,500 men killed and wounded. MacMahon pulled back to a better defensive position.

Breaking through the border

The Prussians did not allow MacMahon to build up his strength, however. On August 6, the Prussian Third Army (125,000 men) attacked at Fröschwiller (also known as Wörth). The outnumbered French (45,000 men) fought stubbornly against the Prussians who were trying to surround them. Heroic charges by MacMahon's cavalry held up the Prussians, but by nightfall the French were in retreat, heading for Châlons-sur-Marne, which they reached on the 14th. French casualties amounted to 17,000, while the Prussians had about 10,000. The Prussian victory had opened the way to Paris.

The Prussian First and Second Armies had been ordered to attack Bazaine's Army of Lorraine. The Army of Lorraine was dispersed into three widely separated groups—if one was attacked the other two would not be able to rush to its aid. The first

Troops from the German state of Württemberg, one of Prussia's allies during the war, attack the French during the Battle of Fröschwiller on August 6, 1870.

THE CHASSEPOT RIFLE

The French army in the Franco-Prussian War was equipped with probably the best rifle of the period. The weapon, the Chassepot, was named after the French official who developed it. The weapon far outclassed the rifle carried by the Prussians, which was called the Dreyse.

The Chassepot had a range of over 4,000 feet (1,200 m). The Chassepot had a better firing mechanism than the Dreyse, which allowed French troops to fire more rapidly than their Prussian counterparts. It also fired a lighter bullet, so French troops could carry more rounds into action than Prussian soldiers.

Despite all of these advantages, the Chassepot proved not to be a battle-winner for the French. It did have problems. The barrel soon became fouled with gunpowder and had to be cleaned frequently, its ammunition often deteriorated quickly especially in damp conditions, and the weapon had a vicious recoil when fired, thereby making accurate fire difficult.

However, the major problems were not the fault of the weapon but rather those who used it. Many French troops were poor marksmen and some had never used the rifle before the war.

French forces to be attacked by the Prussians were the 30,000 men under General Charles Frossard holding the high ground at Spicheren, again close to the border. On August 6, the Prussians tried to surround the French as more and more Prussian troops arrived, building up to 45,000 men. Bazaine was unwilling to support Frossard and the French were forced to withdraw after suffering 3,000 casualties. The Prussians had 4,500.

Moltke was rightly pleased with the way the campaign was developing; his armies had won several battles and the French were in retreat. Moltke ordered his armies to push deeper into France. The Prussian Third Army chased after MacMahon, while the First and Second Armies pursued Bazaine. Their aim was to prevent Bazaine's and MacMahon's armies from linking up.

The Prussian Second Army caught up with Bazaine on the 16th at Mars-la-Tour. Without waiting for the bulk of his army to reach the battlefield, the commander of the Second Army, Prince Friedrich Karl, attacked. He knew the rest of the army would hear the fighting and rush to his aid. The cavalry of both sides launched a series of ferocious assaults against each other in the afternoon. It was one of the last great cavalry actions. Friedrich Karl's other units gradually reached the battlefield and he felt able to order an all-out onslaught against the French. By nightfall both sides had suffered around 17,000 casualties and both camped on the battlefield.

Bazaine did not expect help to arrive, so he withdrew his 115,000 men toward the fortress of Metz. However, this maneuver left Bazaine's army facing in the direction of Paris with its back to the French border. The Prussians lay between him and the French capital. He was cut off. Moltke now ordered that Bazaine's army be forced from the high ground it was occupying between the villages of Gravelotte and St. Privat to the west of Metz and then sealed up inside the city. The two villages became the focus of the fighting.

A French general prepares to order his forces to attack the Prussians at Gravelotte–St. Privat on August 18, 1870.

On August 18, Moltke ordered an attack on Bazaine's forces between Gravelotte and St. Privat. The Prussian Second Army on the left began the assault. The village of St. Privat was the focal point of the first round of fighting. The Prussians launched over 10,000 men against the French garrison of 23,000 from early morning to nightfall. The French were finally forced to retreat from St. Privat.

French forces in disarray

While the battle for St. Privat was raging, a second battle was being fought for Gravelotte on the Prussian right. The Prussian attack became bogged down in a ravine and came close to collapse when the French launched a counterattack. This was only beaten off by massed artillery fire. By midnight Moltke knew that he had won a victory. It had been a hard-fought battle but by the end of the battle Bazaine had been bottled up in Metz. The city was placed under siege.

With Bazaine sealed up in Metz, the Prussians moved to deal with MacMahon's forces. MacMahon had been idling at Châlons since August 14 and was finally ordered to march to Bazaine's aid on the 21st. His army, accompanied by Emperor Napoleon III, moved north, hoping to then swing east and head for Metz. This maneuver left him wide open to attack. Moltke ordered the Prussian First Army and part of the Second to keep up the pressure on Metz, while sending the Third Army and the rest of the Second, known as the Army of the Meuse, after MacMahon.

The Army of the Meuse fought two battles against MacMahon's forces—at Nouart on August 29 and at Beaumont a day later. These battles forced MacMahon farther north, away from Metz. A third battle, at Bazeilles on the 31st, saw the

THE REFFYE MITRAILLEUSE

The Reffye mitrailleuse was the world's first really practical machine gun. Developed in supposedly great secrecy by the French in the 1860s (the Prussians actually knew of its existence), it was built by an army officer, Verchère de Reffye, and 200 had been supplied when production ended in 1868. The weapon was mounted on a two-wheeled carriage and consisted of five banks of five barrels. Trials with the weapon showed that it could fire up to 125 rounds a minute.

The French had many available during the Franco-Prussian War, but the weapons did not perform overly well. There were three main reasons for this. First, the weapon had been developed in secret so few troops had been trained to use it. Officers who had been trained to use it were sent to conventional artillery units at the outbreak of war due to an error in their orders. Second, the weapons were not used properly on the battlefield: Rather than being deployed with the infantry, they were used as a type of conventional artillery. However, they were easily outranged by the Prussian artillery and blown to pieces. Third, they were not mechanically reliable and often unable to stand up to the rigors of campaign.

Despite these shortcomings, the mitrailleuse was a pointer to the future. By the beginning of World War I, the machine gun had become one of the most important weapons available to the frontline soldier.

The mitrailleuse machine gun was no match for the longer-ranged Prussian artillery.

French army pushed into a wide bend in the Meuse River at Sedan. Once again the Prussians had been able to cut off a large French army from Paris.

The Battle of Sedan on September 1 was a catastrophe for the French. MacMahon, who had been wounded at Bazeilles, was replaced by General Auguste Ducrot. Ducrot was in an almost impossible situation—his army had the Meuse at its back and was facing over 200,000 Prussian troops advancing from the north, west, and south. The French had no choice. They had to break out of the trap or they would have to surrender.

Ducrot launched his cavalrymen against the Prussians but they were shattered by rifle fire. Prussian artillery, over 400 cannon, pounded Sedan from the high ground circling the city. German cavalry was thrown back by French machine-gun fire, but time was running out for the French. By late afternoon the French abandoned all hope of breaking out and fled into Sedan.

The Battle of Sedan was over but Ducrot's replacement, Emmanuel de Wimpffen, attempted to persuade Napoleon to lead one last attack against the Prussians. Napoleon knew that it would be pointless and refused. Napoleon left the walls of Sedan under a white flag and surrendered to the Prussian king, Wilhelm I. Wimpffen followed, leading 83,000 French soldiers and over 400 cannon into captivity. Over 15,000 Frenchmen were killed, wounded, or missing. The Prussians had 9,000 casualties.

France fights on

With one half of the French army defeated at Sedan and the other half under siege at Metz, the way was open to Paris. The war was seemingly over. The Prussians believed they had only to march into Paris and subdue Metz and a number of other fortresses to achieve total victory. However, France was swept by a wave of patriotic enthusiasm. Their emperor might be in captivity but the French people were ready to fight on.

In Paris, a popular uprising saw the emperor's government overthrown. It was replaced by the Third Republic. The republic's president, General Jules Trochu, decided to defend Paris. He hastily mobilized what was left of France's armed forces. Some 120,000 ex-soldiers were gathered along with 80,000 *gardes mobiles* (young recruits) and 300,000 *gardes nationales* (older reserves). These were no match for the Prussians in battle but manning Paris's extensive fortifications they were a powerful force.

Moltke had no intention of launching costly assaults against Paris's defenses and ordered his forces to lay siege to the capital. Siege artillery was slowly brought up instead. He believed that either bombardment or hunger and disease would force Paris to surrender. The Prussians were, however, stretched dangerously thin. They were engaged in two large sieges (Metz

A Prussian lancer clashes with a French cavalryman during a skirmish in 1870.

Prussian gunners and their artillery pictured during the siege of Paris, which lasted from September 1870 until January 28, 1871.

and Paris), their lines of communication with Prussia were under attack from bands of French *francs-tireurs* (guerrillas), and the Third Republic government, based in the city of Tours, was raising new field armies in the French provinces.

There was some good news for the Prussians, however. On October 27, Metz fell. Bazaine and more than 170,000 French troops surrendered after enduring a siege lasting 54 days. Moltke moved swiftly. Those Prussians forces freed by the surrender of Metz moved into the valleys of the Loire River and its tributaries to take on the new French Army of the Loire, which was attempting to march to the relief of Paris. Operations were inconclusive and dragged on throughout the winter. The Prussians were also finding it difficult to capture Paris.

Battles outside Paris

On November 9 the French won the Battle of Coulmiers, which forced the Prussians to withdraw from the city of Orléans to the southeast of Paris. The Prussians responded quickly by retaking Orléans on December 4, but another of France's new armies was heading to the relief of the fortress of Belfort. In northern France a French army led by General Louis Faidherbe was also causing problems for the Prussian invaders. Faidherbe fought a drawn battle with the Prussians at Halluin on December 23 and at the two-day Battle of Bapaume on January 2–3, 1871. Although he was defeated at St. Quentin on the 19th, his army was able to escape to fight another day.

The French relief army heading for Belfort arrived outside the fortress in the middle of February. Although the French army consisted of 150,000 men it was woefully inexperienced and was defeated by the 60,000 Prussian besiegers after a three-day battle on January 15–17. The Prussians rushed a relief army to Belfort and forced the French south toward Switzerland. Pinned against the border, the French commander had no option but to cross into Switzerland with 83,000 men. They were held by the Swiss.

On January 26, 1871, the Paris garrison made one final attack to smash through the Prussian siege lines. This failed and Trochu agreed to an armistice with the Prussians. Paris surrendered to the Prussians on January 28. The regular French troops and *gardes mobiles* in the capital were made prisoner and the city's forts were turned over to the Prussians.

The *gardes nationales* were allowed to keep their weapons as the government of the Third Republic needed an armed presence in Paris to prevent outbreaks of lawlessness. Only Belfort continued to resist. In a superb example of defensive warfare, the garrison commander, Colonel Pierre Denfert-Rochereau, held out until February 15. He was finally ordered to surrender by the French authorities. The siege had lasted for over 100 days. The defense of Belfort was the only bright spot in a war that had gone badly wrong for the French.

Soldiers of the French relief force trying to reach the defenders of the city of Belfort are forced to retreat by the Prussians after a three-day battle in January 1871. Belfort, however, continued to resist the Prussians until the middle of the next month.

The Treaty of Frankfurt that ended the Franco-Prussian War was harsh. France was forced to give up to Prussia its border provinces of Alsace and Lorraine, and pay a huge sum of money to the Prussians. The Prussians maintained an army of occupation in France until the money was paid. They did not complete their evacuation until September 1883.

The Paris Commune

The end of the Franco–Prussian War did not, however, end France's agonies. After its capture Paris was the scene of rebellion. Members of the *gardes nationales* overthrew the government of the Third Republic. Between March 18 and 28, 1871, Paris was the scene of great violence and destruction as political rivals sought to settle old scores. A new government—called the Commune—was declared. The old government fled and prepared the regular army—soldiers from the armies captured at Metz and Sedan and released by the Prussians—to recapture Paris.

These loyal French forces began their attacks on Paris in early April but progress was slow as the Communards fought for every inch of ground. The loyal French eventually broke into Paris itself but had to fight for every barricade and fortified house that the Communards occupied. Atrocities and reprisals were commonplace as Frenchman fought Frenchman. It is believed that when the fighting ended on May 28 some 20,000 Parisians had been executed or murdered.

The Franco-Prussian War left Prussia as the head of a newly unified Germany, which had become the most powerful state in Europe. It also left a legacy of hatred between France and Prussia. In little more than 40 years the two would again be at war, but this time the war would be part of a global conflict.

FRANCS-TIREURS

The French *francs-tireurs* were originally groups of volunteers who received some military training in peacetime. They were trained to gather intelligence, ambush enemy patrols, and launch raids against enemy supply lines. When the Franco-Prussian War began in 1870, the ranks of the *francs-tireurs* were swelled by enthusiastic civilians.

The importance of the *francs-tireurs* varied. In some areas of France their activities tied down Prussian troops. Elsewhere, their presence was almost unknown. While some units were filled with skilled, almost professional soldiers, others contained men who acted like thieves and bandits.

The Prussians' reaction to the *francs-tireurs* was equally mixed. In some cases they treated those they captured as criminals rather than prisoners of war and shot them out of hand. However, some *francs-tireurs* were so ineffective that they had little impact on the Prussians or their war plans.

If the *francs-tireurs* were effective in a particular area, the Prussians often vented their anger on the local French civilian population, burning their homes and destroying their crops and livestock in revenge.

NAVAL WARFARE TRANSFORMED

Naval warfare underwent a revolution in the second half of the 19th century. New construction materials, various types of mechanical power, and devastating new weapons heralded the end of the wooden, sail-powered warships of old. New types of warships were also being introduced, which by the turn of the century threatened the role of the large battleships. However, there were no major battles between fleets for the 100 years after the Battle of Trafalgar in 1805, a period during which Britain's naval power was unrivaled.

The British Warrior *(left), launched in 1860, was the first ironclad battleship and the most powerful warship afloat when it was launched.*

The key change in warships in the second half of the century was from the old wooden, sail-powered ship of the line (large warships with many cannon) to the iron, later steel, steam-driven battleship. The process did not occur overnight but was gradual. Much design work was experimental and many ships were unique, such as the circular *Admiral Popov* and *Novgorod* built by Russia, for example.

The Devastation, *a warship of the British navy, was launched in 1871. It was the first vessel to completely do away with sails. Power was provided by coal-fired engines.*

At first the old wooden ships of the line were simply covered in iron sheeting, and kept their sails. Next, all-iron ships were produced with steam engines, although they often carried some sails. Then the sails were abandoned as the engines became more reliable. The first true ironclad battleship was the British navy's *Warrior* launched in 1860. The first battleship to drop all sail power was the British *Devastation*, designed in 1869.

New ships and weapons

The transition from sail to steam meant that a ship's range (the distance it could travel without refueling) was restricted to how much coal (and later oil) it could carry. Consequently, major powers with large navies had to acquire refueling stations at strategic points around the world. Britain, the leading sea power in the 19th century, struck deals with local rulers or established major naval bases in British colonies where warships could take on coal. These included places such as Singapore and Trincomalee, in what is now Sri Lanka. Other global powers with large navies did the same thing.

The new battleships also had a different layout for their weapons. The old wooden ships of the line had cannon in broadsides (row above row) on either side of the vessel, as did many of the first ironclad transitional vessels. The true revolution came with the revolving turret, which was first used in action by the

THE BATTLE OF NAVARINO

Navarino in 1827 was the last large naval battle in which wooden, sail-powered warships blasted it out at close range. A British, French, and Russian fleet went to the aid of Greek nationalists who were trying to evict the Turks from Greece and arrived at Navarino Bay on the southwest coast of Greece on September 11. On October 20, the commander of the outnumbered allied fleet, Admiral Sir Edward Codrington, ordered his warships into the narrow confines of the bay to prevent the Turkish fleet from putting to sea.

The Turks, led by Ibrahim Pasha, opened fire on a pair of small allied craft and the allied gunners immediately opened fire in return. The action was intense but increasingly onesided as the allied ships blasted the Turkish vessels to pieces at close range. After two and a half hours Ibrahim's force had been destroyed. None of the allied vessels was sunk, although several, chiefly British, were damaged.

U.S.S. *Monitor* at the Battle of Hampton Roads in 1862 during the American Civil War. The turret was a usually enclosed compartment on the deck of a vessel that contained one or two guns. The turret could revolve, allowing the guns to fire in virtually any direction. The turret was gradually adopted by the new iron and steel battleships. Battleships usually had two turrets, generally positioned on the front and rear decks, although some had them mounted in the center of the deck.

There were also new developments in naval gunnery. The old muzzle-loading cannon were gradually replaced by new breech-loading types. Gunpowder was replaced by the more reliable cordite charge (a smokeless explosive made from cellulose nitrate and nitroglycerine). Cannonballs—useless against iron or steel, which they just bounced off—were replaced by the explosive shells shaped like a cone that could penetrate iron and steel.

Although the newer battleships had far fewer guns than the old ships of the line, those they had were more reliable and accurate, and had a longer range than the older guns. Naval battles no longer had to take place at just a few hundred yards or even less but could be fought out over much greater distances.

The first torpedoes

The late 19th century also saw the gradual introduction of a new and potentially devastating weapon—the underwater torpedo. The best of a number of torpedo designers was an Italian-based Englishman, Robert Whitehead, whose first design was built in 1866. These early torpedoes were fired from surface vessels. The first self-propelled torpedo was used in action in 1877 when the British warship *Shah* attacked but failed to sink the Peruvian *Huascar*, whose crew had mutinied and was attacking British vessels off the coast of South America in random acts of piracy.

The torpedo was also responsible for the development of a new type of small surface warship—the torpedo boat. These speedy vessels carried a number of torpedo tubes and were considered such a threat to the bigger battleships that a new warship was developed to counter them. This vessel was the destroyer (originally torpedo boat–destroyer), a fast, gun- and torpedo-armed warship. The first, Britain's *Havock*, was launched in 1893. It carried torpedoes and guns, and had a high top speed.

However, the torpedo really came into its own with the development of the submarine. The first successful submarines were designed by an Irish-born American, J.P. Holland, and these "Holland" boats were used by the U.S. and British navies.

Key technical developments at the end of the 19th century made the modern submarine possible. First was the development of an internal-combustion engine that was small enough to be fitted into a submarine and was used to power it on the surface. Second was the electric motor powered by batteries, which allowed the submarine to travel underwater and did not require a supply of air to work.

In less than 50 years naval warfare and warship design had been transformed. By 1900 navies had a greater variety of more powerful ships than ever before. These ships were used to fight wars around the globe and gave major powers a world role.

A U.S. torpedo boat–destroyer (right) launches an attack on a Spanish ship during the Spanish–American War of 1898. These small, speedy ships proved highly successful in using their torpedoes to sink larger vessels.

Native American Wars

From the end of the American Civil War in 1865 until the 1890s, there were almost continuous clashes between Native Americans and the U.S. Army, which was protecting white settlers eager to colonize Native American lands in the West. The wars usually consisted of hit-and-run raids by the Native Americans, who were then chased by U.S. cavalry columns that operated out of a number of forts. The Native Americans won some battles but could never hope to win outright victory against the might of the United States.

U.S. soldiers and civilians use their repeating rifles to beat off a large Sioux war party led by Red Cloud during the Hayfield Fight on August 2, 1867.

As soon as the U.S. Army returned to the West after the Civil War, it became embroiled in a campaign in Wyoming and southern Montana in 1866–67. Sioux, Cheyenne, and Arapaho opposed the building of new forts along the Bozeman Trail, which was to be used to further open up the Rocky Mountains and beyond for white settlers eager to exploit the gold fields of Colorado. The trail passed through lands that had been reserved for the Native Americans by treaty. The fighting, which lasted for

NATIVE AMERICANS AT WAR

The traditional weapons carried by the Native Americans were the bow and arrow, long lance, club, tomahawk (a length of wood topped with stone or ax-head) and knife. Both metal tomahawks and knives were often bartered with Europeans in exchange for hides. Native Americans might also carry shields for protection.

As the Native Americans came into contact with white traders, they began to acquire a range of firearms. These were often old weapons, but Native Americans fought with shotguns, hunting rifles, revolvers, and military carbines. Some weapons were taken from the dead after a battle or skirmish, while others were actually given to Native Americans by the U.S. government for hunting.

Most Native Americans in the Midwest, Rockies, and Southwest fought on horseback, although they did sometimes fight on foot. The Apache of the Southwest used horses to get close to an enemy but did their actual fighting on foot.

a year, became known as Red Cloud's War, after a Sioux chief. Red Cloud besieged Forts Reno, Phil Kearny, and C.F. Smith, and then annihilated a small cavalry force led by Captain William Fetterman near Fort Phil Kearny on December 21, 1866 (see map page 44).

On August 1, 1867, about 500 Cheyenne attacked 30 soldiers and civilians a little way from Fort C.F. Smith. This battle, the Hayfield Fight, was very one sided. The soldiers were armed with new repeating rifles and protected by a log corral. The Cheyenne, with 20 warriors dead and many injured, could not get near the corral because of the intensity of the soldiers' fire The Cheyenne war party finally retreated under the cover of the flames and smoke from the dry grass they had set ablaze. The next day a Sioux war party led by Red Cloud attacked woodcutters near Fort Phil Kearny. The soldiers with the woodcutters also carried repeating rifles and this battle, known as the Wagon Box Fight because the woodcutters took shelter behind their wagons, ended as did the Hayfield Fight.

Red Cloud's War ended with the Treaty of Fort Laramie, which was signed in April 1868. Red Cloud promised not to go to war again and the U.S. authorities abandoned their forts. The territory to the north of the North Platte River through which the Bozeman Trail was originally supposed to pass was forbidden to white settlers.

U.S. troops take shelter on Beecher's Island, Colorado, as they try to beat off attacks by Cheyenne warriors led by Roman Nose in September 1868. The Cheyenne kept the troops under siege for over a week before they were forced to withdraw by the arrival of an army relief force.

The U.S. Army began campaigning in the Colorado Territory in 1868 as friction between white settlers, gold prospectors, and Native Americans grew. In early fall Major George Forsyth fought the Cheyenne. Lieutenant Colonel George Armstrong Custer also led his Seventh Cavalry against the Cheyenne. Forsyth's party was attacked by about 500 warriors. The U.S. troops took up position on a rise in the middle of a dried-up riverbed, where they were besieged by the Cheyenne for eight days (September 18–25). This battle, called Beecher's Island, ended when a relief column reached Forsyth.

In November Custer caught up with some of those Cheyenne who had fought at Beecher's Island. They had traveled to their homeland along the Washita River in present-day Oklahoma. On the 27th, Custer charged out of a fog. When the fighting ended, 103 Cheyenne, mainly women and children, were dead.

The Little Bighorn campaign

In the mid-1870s, settlers were pushing into the Dakota Territory, where gold had been found in the Black Hills. This was an area sacred to the Sioux and to many Cheyenne, and had been reserved for them by the treaty that had ended Red Cloud's War. The uprising against the settlers was led by Chief Crazy Horse of the Oglala Sioux and Chief Sitting Bull of the Hunkpapa Sioux.

As the uprising spread, the U.S. authorities sent General George Crook and 800 cavalrymen to the area in early 1876. Crook surprised Crazy Horse at Slim Buttes, his winter camp, on the Powder River on March 17. The fight at first went well for Crook but he was eventually forced to retreat.

Control of the campaign passed to General Alfred Terry, who sent two other columns of U.S. soldiers into the Black Hills to link up with Crook on the Yellowstone River. Crook's column stumbled into Crazy Horse, who had about 5,000 warriors with him, on June 17. Crook was outnumbered by five-to-one but the outcome of the Battle of the Rosebud, a creek flowing into the Yellowstone River, was undecided. Crazy Horse withdrew to avoid being trapped, while Crook fell back to reorganize his battered command. Terry, however, knew nothing of Crook's plight or the Battle of the Rosebud and continued to pursue Crazy Horse, sending General George Custer's Seventh Cavalry, some 600 men, south to get behind Crazy Horse.

Custer, a veteran of the American Civil War, and considered by some to be a glory-seeking commander, caught up with Crazy Horse along the Little Bighorn River. Custer should have waited for the rest of Terry's forces to catch up with him, but decided to attack on June 25. Custer made a fateful error of judgment—he divided his small command into three groups in the face of a much larger enemy. Custer led 212 of his men into the center of Crazy Horse's forces and was wiped out, and the two other

A romanticized version of Custer's (top) last stand during the Battle of the Little Bighorn. on June 25, 1876. Custer was responsible for his own defeat as he made the error of dividing his regiment in the face of a stronger enemy.

columns had to fight alone for two days until Terry arrived. The Battle of the Little Bighorn was the greatest disaster suffered by the U.S. Army in all its wars with the Native Americans.

The U.S. forces were unable to trap Crazy Horse and his followers for several months, but there was one major fight—the Battle of Crazy Woman Fork on the night of November 25–26. Ten troops of cavalry led by Colonel Ranald MacKenzie surprised a Native American camp and destroyed it. It was not until

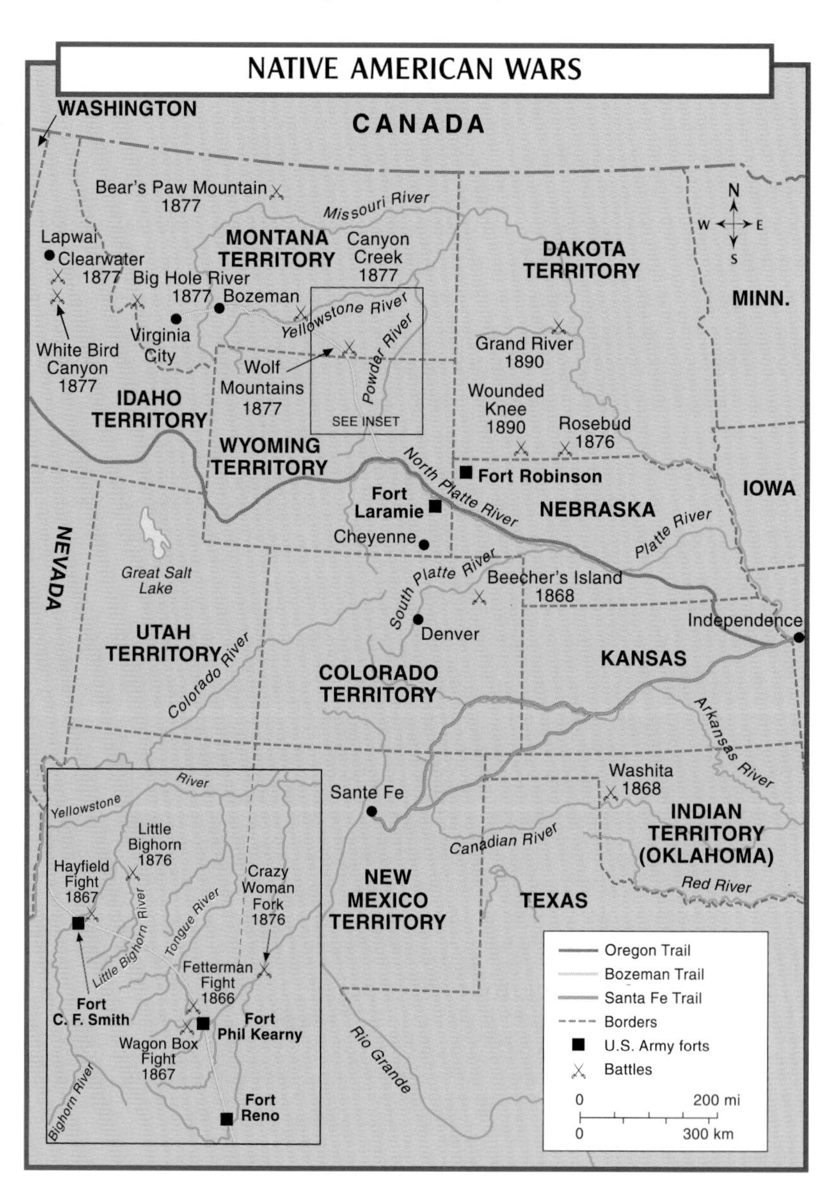

The major battles fought between the U.S. Army and the Native Americans in the second half of the 19th century.

January 1877 that U.S. forces, 500 infantrymen and two cannon under Colonel Nelson Miles, were able to track down Crazy Horse. The Battle of Wolf Mountains on the 7th ended with Miles's troops destroying Crazy Horse's village. Crazy Horse was finally forced to surrender at Fort Robinson, Nebraska, on May 6. Crazy Horse was arrested for trying to lead another Sioux uprising in September. He was killed, while trying to escape, on the 6th.

Chief Joseph of the Nez Percé is widely regarded as the greatest Native American military leader.

The war of Chief Joseph

In 1877, the U.S. authorities had also to deal with the Nez Percé under Chief Joseph, probably the greatest Native American military leader. The Nez Percé were ordered to leave their homeland in Oregon for a reservation near Lapwai (now Lewiston), Idaho, assigned to them, and Chief Joseph agreed under pressure from the U.S. Army in June. However, some white settlers were killed by a group of Chief Joseph's warriors and a cavalry detachment was sent by General Oliver Howard to attack the Nez Percé. Chief Joseph and his warriors fought back with great skill, virtually destroying the cavalry detachment at the Battle of White Bird Canyon on the 17th.

Pursued by U.S. Army troops Chief Joseph now led his tribe (just 700 people, including 300 warriors) on a great trek of nearly 2,000 miles (3,200 km) that took the Nez Percé through Oregon, Idaho, Wyoming, and Montana. Chief Joseph was trying to lead his people to safety in Canada.

For four months, as he headed for Canada, Chief Joseph outthought and outfought the U.S. commanders who were pursuing him and his people. He defeated Howard at the Battle of the Clearwater (July 11–12) in Idaho and then won victories at the Battles of Big Hole River (August 9–10) and Canyon Creek

(September 13), both in Montana. Chief Joseph's victories allowed him to use captured U.S. weapons and supplies to equip and feed his people. Joseph was just 30 miles (48 km) from Canada, at Bear's Paw Mountain, when he was caught by General Nelson Miles on September 30. Joseph fought for four days, despite being outnumbered, but was forced to surrender.

Two of the U.S. officers, Howard and Miles, who had chased Chief Joseph and the Nez Percé over such long distances did, to their credit, try to get them returned to their homeland in Oregon. However, higher authorities decreed that the Nez Percé had to be sent to the Indian Territory (now Oklahoma).

The final campaigns

In the 1880s the U.S. Army launched a campaign against the Apache in Arizona and New Mexico. The key Apache leader was Geronimo, who saw that he could not hope to defeat the U.S. Army in a conventional battle with the number of warriors at his disposal. Geronimo decided to use classic hit-and-run tactics to strike at places where the army was weakest and avoid large detachments of U.S. forces.

Geronimo raided in Arizona and New Mexico until he was forced to surrender by General George Crook on two occasions. Geronimo escaped both times. After the second escape in March

U.S. troops survey the aftermath of the onesided fight at Wounded Knee, South Dakota, December 29, 1890. Wounded Knee was the last major act of the U.S. Army's fight against the Native Americans.

1886, the U.S. Army sent General Nelson Miles to deal with Geronimo. Geronimo had few warriors to protect more than 100 women and children. Miles had some 5,000 troops and 500 Native American scouts under his command.

Despite the odds, Geronimo defeated every attempt by Miles to capture him. For over five months, he crisscrossed Arizona and New Mexico, always one jump ahead of the U.S. Army. Geronimo was never completely defeated but chose to surrender on September 4, 1886. Geronimo was placed in prison in Florida before he was allowed to settle in Oklahoma in 1894.

Wounded Knee

By 1890, the U.S. Army had gained the upper hand against the Native American tribes, but there was one last, tragic act to play out. On December 15, Sitting Bull was killed in a battle on the Grand River, South Dakota. The leadership of the Sioux now passed to a chief named Big Foot.

The U.S. Army sent Colonel James Forsyth and the Seventh Cavalry, Custer's old unit, to return Big Foot and his followers to their reservation. Forsyth caught up with Big Foot at Wounded Knee, South Dakota. When the troopers tried to disarm Big Foot's followers, firing broke out. When the firing stopped, Big Foot and an estimated 200 of his followers, including women and children, were dead. The cavalry suffered 25 killed and 39 wounded.

This "battle" at Wounded Knee was the last major event of the long, destructive wars that had seen the Native American tribes of the United States surrender their tribal homelands, and often lose their traditional ways of life and culture, in the face of the steady westward expansion of white settlers eager to colonize and exploit new lands.

GHOST DANCE

The Ghost Dance was a Native American religious cult that was begun in the later 1870s by Wovoka, a Paiute shaman (religious leader). Its central rite was a dance that lasted five nights. The religion promised reunion with the dead, eternal life, peace, and a Native American world free of white interference. The new religion spread rapidly at a time when the fortunes of the Native Americans were at a low ebb after major defeats at the hands of the U.S. Army and the loss of homelands to white settlers.

The Sioux were particularly attracted to the Ghost Dance religion and Chief Sitting Bull developed Wovoka's teaching further, believing that war with the white authorities and settlers was inevitable. The U.S. government, fearful of more unrest among the Native Americans, attempted to destroy the religion and capture its leaders. Sitting Bull was arrested and killed when he attempted to escape on December 15, 1890. The U.S. Army eventually caught up with a large party of Sioux believers at Wounded Knee, South Dakota, on the 29th. Their defeat in the "battle" that followed effectively ended the Ghost Dance religion, and Native American resistance to the U.S. government.

THE ZULU WAR

The British fought a great number of wars and campaigns to expand their empire during the second half of the 19th century. One of the most remarkable—and the one in which they came closest to disaster—was the Zulu War. The Zulus were a warrior people in southern Africa who had carved out an empire in the region under their king Shaka (1787–1828). As the British pushed inland from their colonies on the coast, they came into contact with the Zulus and then stirred up a war in order to take over the Zulu lands.

On December 11, 1878, the British colonial authorities in southern Africa issued a series of demands to Cetewayo, the Zulu king. To agree to these demands would have meant that the Zulus would have had to give up their independence. Cetewayo refused outright. He had no desire to go to war but was determined to protect his lands from invasion.

The rival armies

The British assembled an invasion force of 5,000 British regular troops and local Europeans, and more than 8,000 local native conscripts under Lord Frederick Chelmsford. Chelmsford was short of cavalry and wagons to carry his supplies but decided to invade Zululand at three points. Like many British of the time he felt that a modern British army equipped with cannon and rifles would be more than a match for his opponent.

In fact the Zulus were formidable warriors. Armed with thrusting and throwing spears, shields, and wooden clubs, they

A Zulu warrior dressed in ceremonial clothing. His clothes would have been much simpler during wartime.

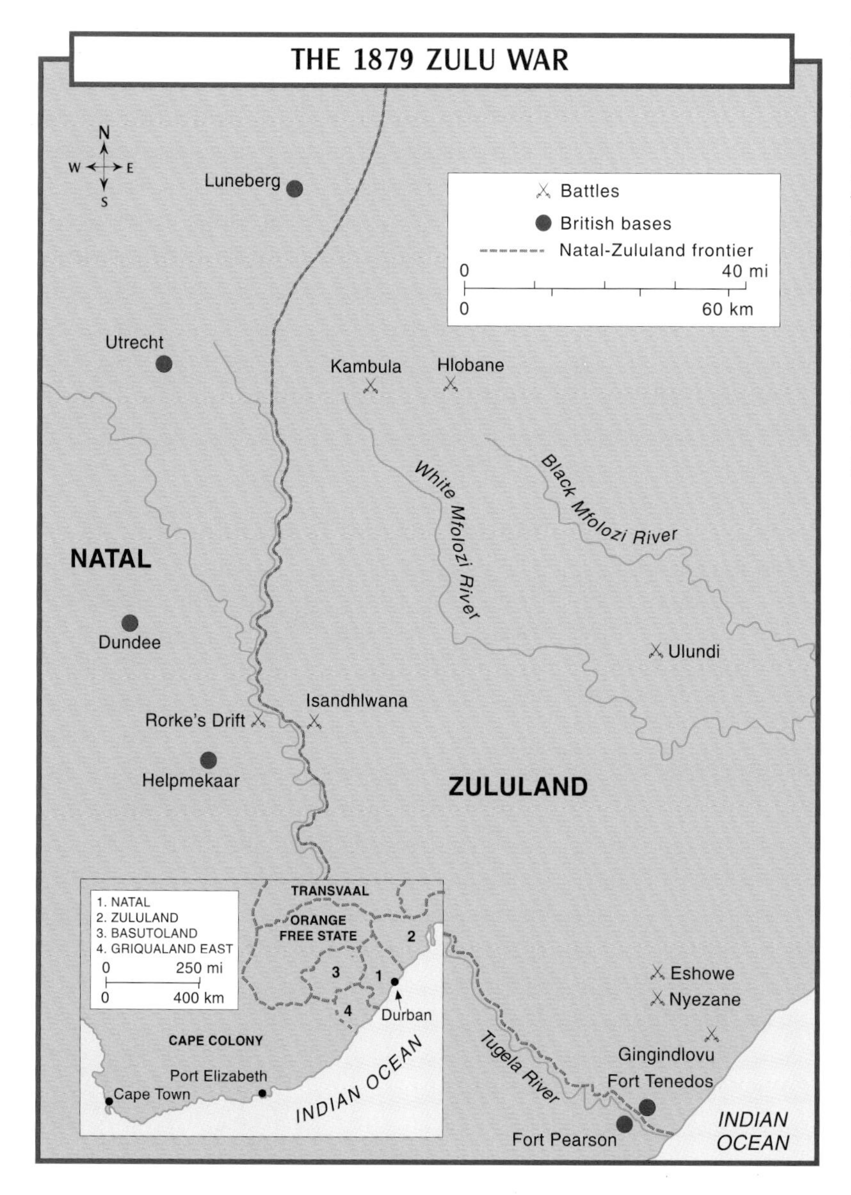

The Zulu War was begun by the British who wanted to take over Zululand. The Zulus fought back and defeated the first British invasion. However, the second invasion saw the British win a major victory at Ulundi, which destroyed Zulu military power.

were skilled fighters at close range. The Zulus were also highly organized; Cetewayo had an army of 40,000 men divided into a number of regiments. These could cover many miles a day on foot and had a well-tried and successful battle plan. A Zulu army on sighting an enemy formed up in four parts. The center block of the army attacked the enemy head-on, while the two blocks on the wings would fan out and encircle the enemy. The final block was held in reserve, ready to finish off the enemy if needed.

THE ZULU ARMY

The Zulus had a well-organized military system. The basic military unit, the impi, was roughly equivalent to a regiment, which could vary in strength from around 500 men to more than 5,000. The average impi contained about 1,500 warriors. The impis were made up of warriors of the same age.

Each regiment had a distinctive ceremonial uniform and carried an oval shield made from cowhide. Younger warriors tended to have mainly black shields, while older ones had white or reddish ones.

The main Zulu weapons included a thrusting spear with a long, broad blade and a lighter spear for throwing. Some Zulus did have firearms. These were generally outdated weapons. However, the Zulus did capture a large number of modern rifles after crushing a British force at Isandhlwana.

Zulu forces were famed for the speed of their movement. The warriors generally averaged 20 miles (32 km) in a day, although distances of up to 40 miles (64 km) were not unknown. A British force would have been lucky to move more than ten miles (16 km) a day on a campaign. The Zulus traveled light, while the British needed huge numbers of oxen and wagons to move their supplies and ammunition.

The British invasion began on January 11, 1879, but was soon in trouble. The central column with Chelmsford at its head pushed into Zululand. It consisted of around 1,800 European soldiers and 1,000 local native conscripts. After a few days Chelmsford and part of the column pushed on ahead and the remainder (1,700 men) was ordered to set up camp below a hill known as Isandhlwana. The officer in charge failed to build a fortified camp. Early on the morning of the 22nd, 20,000 Zulus attacked.

Slaughter at Isandhlwana

The British grabbed their weapons and formed up some way from the camp and its supply wagons. Their rifle fire mowed broad paths through the Zulus. However, the British began to run out of ammunition. They sent runners back to the camp for bullets. Before they could return, the British front line ran out of ammunition and was overwhelmed by the Zulus. The Zulus raced into the camp, killing most of those present. Less than 60 Europeans and 400 native conscripts escaped.

Zulus from Isandhlwana attacked a British camp at Rorke's Drift. The post was defended by 139 soldiers, 35 of whom were sick. The Zulus had about 3,500 warriors. The British fortified Rorke's Drift, and in a battle lasting from the afternoon of January 22 to the morning of the 23rd, they successfully fought off the Zulus.

This British victory, however, was followed by more defeats. The Zulus defeated a British force at Hlobane on March 28. The southernmost British column, after a victory at Nyezane, built a camp at Eshowe but was besieged. The siege lasted from the end of January until early April, when a relief column defeated the Zulus at Gingindlovu.

British lancers complete the destruction of the Zulu army during the closing stages of the Battle of Ulundi on July 4, 1879.

However, Hlobane was the last great Zulu victory. A day later a British force showed how devastating cannon and rifle fire could be. At Kambula on March 29, 2,700 British, troops took on a Zulu army of 20,000 warriors. The British fought from behind trenches and wagons on high ground. The Zulus attacked and were shot to pieces. At least 1,000 were killed.

The British spent most of April and May of 1879 gathering troops for a second invasion of Zululand. This time nothing was going to be left to chance. Troops were under strict orders to build fortifications when they stopped for the night. By the end of May Chelmsford was ready to invade Zululand again.

A simple, direct plan

The second British invasion was far less complex than the first. Only one column was used—4,500 European troops and 1,000 native conscripts. Chelmsford planned to head directly for the Zulu capital, Ulundi, thereby forcing the Zulus to attack him.

On July 4, the British troops advanced on Ulundi in a square formation, with cannon and machine guns at each corner. The Zulus took the bait and 20,000 attacked. None reached the British. Hundreds were cut down by rifle and artillery fire. As the Zulus wavered, British lancers charged out from the square and routed them. The Zulus had at least 1,500 warriors killed, while the British suffered just 100 casualties. The Battle of Ulundi broke the back of Zulu power. Their lands were made part of the British Empire. Cetewayo himself was captured on August 28.

Britain's Wars in Egypt

Egypt was still part of the Ottoman Empire in the mid-1800s but Egypt's chosen leader, Ismail Pasha, was eager to modernize his country. One of his key projects was the building of the Suez Canal, which linked Europe with the Far East by way of the Mediterranean and Red Seas. The project was backed by French finance and technical expertise, but Ismail kept a 44 percent interest in the canal, which was opened in November 1869. However, Ismail's plans left Egypt in great debt and made the regime very unpopular.

Mahdist warriors mounted on camels advance into battle.

Unrest in Egypt grew into a full-scale rebellion led by Ahmet Arabi in 1881, which threatened France's and Britain's strategic interests. Both feared that their warships and merchant ships might be refused access to the Suez Canal. A combined fleet was sent to Egypt in 1882 and on July 11 this fleet's British warships bombarded the Egyptian port of Alexandria, where Europeans had been massacred by Arabi's supporters during the previous month.

The bombardment was followed by the landing of 25,000 British troops. The British now sought out Arabi's main army. During an attack at Tel el-Kebir on the morning of September 13, Arabi's troops were defeated. The British now took a dominant position in Egyptian affairs.

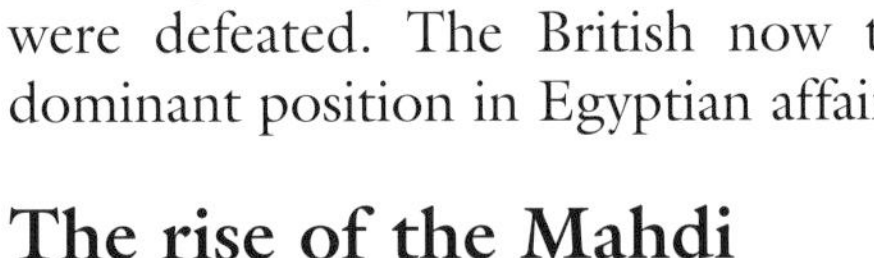

The rise of the Mahdi

While Britain was effectively taking over Egypt, the Egyptian-controlled province to the south, the Sudan, was being swept by a religious-inspired rebellion led by Mohammed Ibn Ahmed el-Sayyid Abdullah, known as the Mahdi (Guided One of the Prophet). In November 1883, the Mahdi's Muslim forces, known as Mahdists, wiped out a British-officered Egyptian military expedition of 7,000 men

THE MAHDISTS

The Mahdists were a strict Muslim sect who followed the teachings of the Mahdi. The Mahdi demanded that his followers lead a simple life of poverty. They were forbidden to drink, use foul language, dance, or take part in festivals. Punishments were severe.

This simple way of life was also reflected in their clothing, which consisted of plain white cotton tunics and short trousers with black patches. The black patches were gradually replaced with more colorful ones. Originally the Mahdists had little military training or organization, but this changed with the Mahdi's successor, Khalifa Abdullah.

The Mahdists were organized into units akin to regiments, which were combined into larger corps. Part of the army was mounted on either camels or horses but most Mahdists were foot soldiers. Their weapons consisted of spears, swords, and shields. However, rifles were acquired through trade or, more importantly, taken from the dead of those armies defeated by the Mahdists in battle. The Mahdists also captured artillery and a few machine guns, but these were used to defend settlements along the Nile River rather than by the main army in the field.

The chief Mahdist tactic involved creeping up on an enemy and then launching a charge at close range. The sudden attack often overwhelmed an enemy. The British, however, tried to counter this by placing their troops behind thorn-tree barricades. If the Mahdists could be kept at a distance, then the superior firepower of the British was almost certain to triumph.

near El Obeid. One of the Mahdi's generals, Osman Digna, crushed a second expedition near Trinkatat in the eastern Sudan. The British, who had virtually no troops in Egypt or the Sudan, ordered the Egyptians to abandon the Sudan.

Gordon at Khartoum

A British officer, Charles Gordon, was ordered to oversee the evacuation of the Sudan and made his base at Khartoum, Sudan's capital, which was located at the point where the Blue and White Nile Rivers meet. Gordon arrived in January 1884 and the Mahdi's forces placed the city under siege. Gordon had few military resources and as the siege dragged on shortages of food and ammunition made his position much worse.

Eventually, the British decided to send a relief expedition, but troops would have to travel a very long way across dangerous desert terrain from Cairo to Khartoum. The only really practical

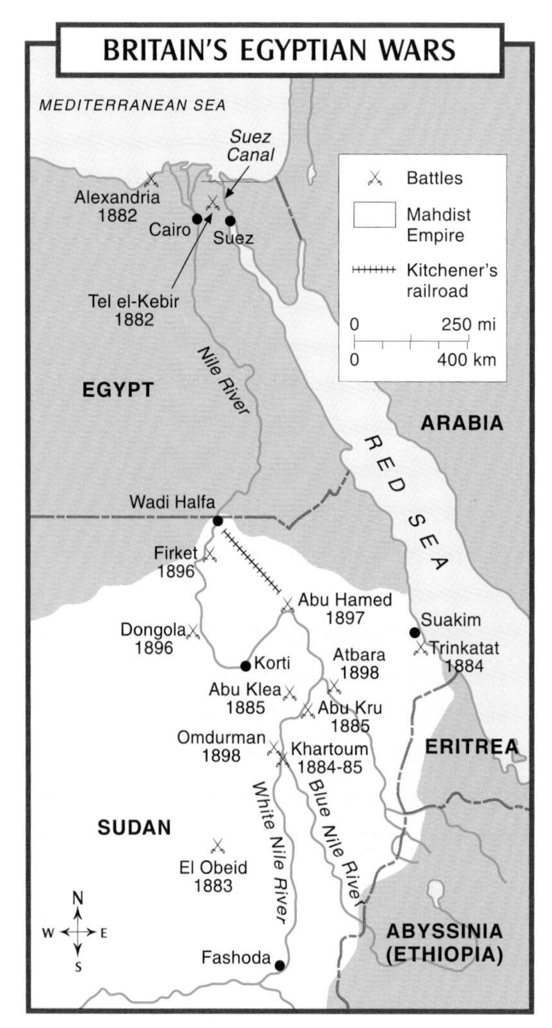

The British fought several campaigns to take effective control of Egypt and the Sudan in the latter part of the 19th century. Victory over the Mahdists at Omdurman in 1898 meant that the British dominated the region.

route to Khartoum was down the Nile River, but the Nile was blocked by dangerous rapids at many points and much of it was uncharted. The relief expedition sailed for Khartoum in October but made very slow progress.

When the expeditionary force was at Korti about 200 miles (320 km) from Khartoum, it was decided to send a fast-moving column of troops overland to Khartoum. This force, known as the Desert Column, began to march in January 1885. The Desert Column beat off two attacks, at Abu Klea on the 17th and at Abu Kru two days later.

On the 21st river steamers from Khartoum reached the Desert Column, clear proof that Gordon was still holding out. After what was to prove a fatal delay of two days, some of the steamers, packed with troops, headed back for Khartoum. They arrived at Khartoum on the 28th. Khartoum, however, had fallen two days before after a siege lasting over 300 days. Gordon had been killed in the final attack. The British withdrew from the Sudan.

The British return

The Mahdi died in June 1885 and the empire he had created was left in the hands of Khalifa Abdullah, one of the Mahdi's generals. Finally, in 1896, the British launched another expedition to reconquer the Sudan. This expedition was commanded by General Herbert Kitchener. His army of 25,000 troops included British, Egyptian, and Sudanese units.

Kitchener planned the campaign in great detail. He ordered a railroad built to transport troops and constructed a number of shallow-draught river steamers that could travel through the Nile rapids. Kitchener made slow but steady progress in 1896, defeating the Khalifa's forces at Firket (June) and Dongola (October). The advance up the Nile continued in 1897, with Kitchener's forces winning a victory at Abu Hamed in August.

By the beginning of 1898, Kitchener was ready to move against the main Mahdist army which was in position around Omdurman, just a few miles away from Khartoum. In April a Mahdist force was crushed at Atbara. By the beginning of September the British were within striking distance of Omdurman. Kitchener placed his troops in a horseshoe formation, with the Nile at their back and a protective thorn barricade (zariba) to the front. The army was supported by machine guns and artillery, and could count on fire support from the river steamers. Kitchener waited for the Mahdists to attack him.

On September 2, the Mahdist forces, some 40,000 men, attacked Kitchener's army. They charged the British position but were swept away in a storm of rifle, machine-gun, and artillery fire. None reached Kitchener's troops; most were cut down hundreds of yards from the British position. Mahdist losses at Omdurman were enormous—10,000 killed, an equal number wounded, and 5,000 captured. Kitchener's casualties were less than 500, and his troops occupied Omdurman after the battle. The Mahdists had been destroyed and Britain's control of Egypt and the Sudan was confirmed.

Supported by fire from gunboats on the Nile, Kitchener's forces blaze away at the Mahdists during the Battle of Omdurman, September 2, 1898.

The Spanish–American War

The Spanish–American War was fought in 1898 and was the largest and most successful overseas operation launched by the U.S. armed forces in the 19th century. The United States had been angered by the way the Spanish had treated their Cuban colonial subjects when they rebelled in the 1890s. Cuban civilians had been evicted from their homes and placed in camps, where many died. However, it was the mysterious sinking of a U.S. warship that finally led to war being declared.

On February 15, 1898, the U.S.S. *Maine* was lying at anchor in Havana harbor, Cuba, when it was blown apart by a sudden explosion. Over 250 of the battleship's crew died. The cause of the explosion has never been fully explained, but U.S. investigators decided that the explosion was not caused by a mechanical problem within the ship. The only other explanation was that an external force—a mine—was the culprit. Spain was blamed.

Certain sections of the U.S. press, particularly newspapers owned by William Randolph Hearst, whipped up anti-Spanish feelings and called for war. Powerful figures in the government were also eager to protect or extend U.S. sugar interests in the

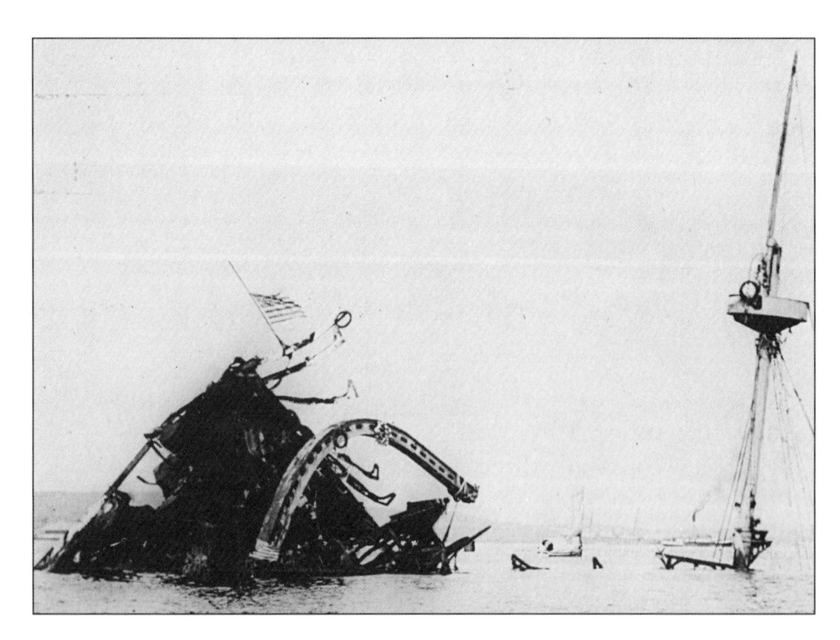

The wreck of the U.S.S. Maine *in Havana harbor. The causes of the battleship's sinking in February 1898 have never been identified. At the time, the Spanish were blamed for the loss and the United States went to war over the event.*

ADMIRAL GEORGE DEWEY

Admiral George Dewey, commander of the U.S. forces at the Battle of Manila Bay, saw active service in the Civil War, during which he served with the U.S. Navy. He took part in the attacks on New Orleans (April 1862) and on Port Hudson (March 1863). After the war Dewey rose through the ranks and was made president of the Board of Inspection and Survey in 1895. A year later Dewey was promoted to the rank of commodore.

In 1897 Dewey requested a return to sea duty and was given the command of the U.S. Navy's Asiatic squadron in November. The Spanish–American War broke out in April 1898 while Dewey's squadron was at anchor in Hong Kong harbor. Dewey was ordered by telegraph to sail for the Spanish-controlled Philippines. He won the Battle of Manila Bay there in early May, destroying the Spanish fleet.

Dewey's crushing victory at Manila Bay was in large part due to his aggressive tactics and intelligent planning. The battle made him a national hero in the United States and he was promoted to the rank of admiral of the navy in March 1899. The rank had been specially created for him. Dewey continued to serve his country as president of the Navy General Board after the war, despite being over the normal retirement age.

region. Although there were voices in government that opposed taking direct military action against the Spanish, the United States declared war against Spain on April 25.

The war had two areas of operations—the Philippines in the Pacific and Cuba in the Caribbean. Both were Spanish colonies. Although the United States had declared war, its understrength army was far from ready to fight. Some 200,000 men volunteered for action but they had to be trained. However, the U.S. Navy was ready for war.

War in the Philippines

The first action of the war took place in the Philippines. When war was declared, Commodore George Dewey, commander of the U.S. Asiatic squadron, was refueling in the harbor at Hong Kong. He ordered his warships—five cruisers and two smaller gunboats—to head directly for Manila, the capital of the Philippines, where there was a small Spanish fleet. Dewey arrived off Manila during the night of April 30, 1898, and ordered his squadron into the confined waters of Manila Bay. The first battle of the war took place the next day.

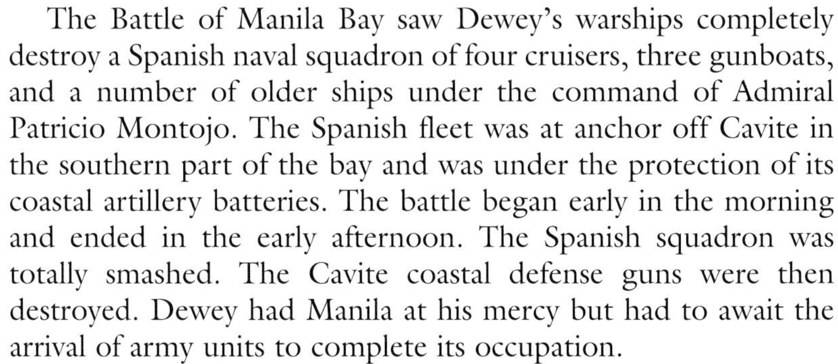

The Battle of Manila Bay saw Dewey's warships completely destroy a Spanish naval squadron of four cruisers, three gunboats, and a number of older ships under the command of Admiral Patricio Montojo. The Spanish fleet was at anchor off Cavite in the southern part of the bay and was under the protection of its coastal artillery batteries. The battle began early in the morning and ended in the early afternoon. The Spanish squadron was totally smashed. The Cavite coastal defense guns were then destroyed. Dewey had Manila at his mercy but had to await the arrival of army units to complete its occupation.

Some 10,000 U.S. troops led by General Wesley Merritt arrived off Manila on June 30 and disembarked at Cavite. Once these troops were ashore, they placed Manila under siege aided by Filipino guerrillas commanded by Emilio Aguinaldo, a leading figure in a rebellion against the Spanish in 1896. There was no hope for the Spanish garrison in Manila. It was too far away from home to expect reinforcements. The U.S. forces attacked on August 13 and the city surrendered.

Commodore George Dewey (center, with white mustache) pictured on his flagship, the Olympia, *at the height of the Battle of Manila Bay on May 1, 1898.*

The war in Cuba

With the Spanish defeated in the Philippines, the focus of the war switched to Cuba. The Spanish had already moved swiftly to build up the island's defenses. On April 29, Admiral Pascual Cervera had sailed from the Cape Verde Islands, a Spanish colony off the west coast of Africa, at the head of four cruisers and three destroyers. Cervera was able to avoid the U.S. naval blockade around Cuba. His fleet anchored at Santiago on the south coast of the island on May 19.

Although surprised by Cervera's arrival, the commander of the nearby U.S. squadron, Rear Admiral William Sampson, reacted quickly. He sailed for Santiago and bottled up the Spanish in its harbor. The Spanish ships could not prevent the landing of U.S. ground forces on Cuba.

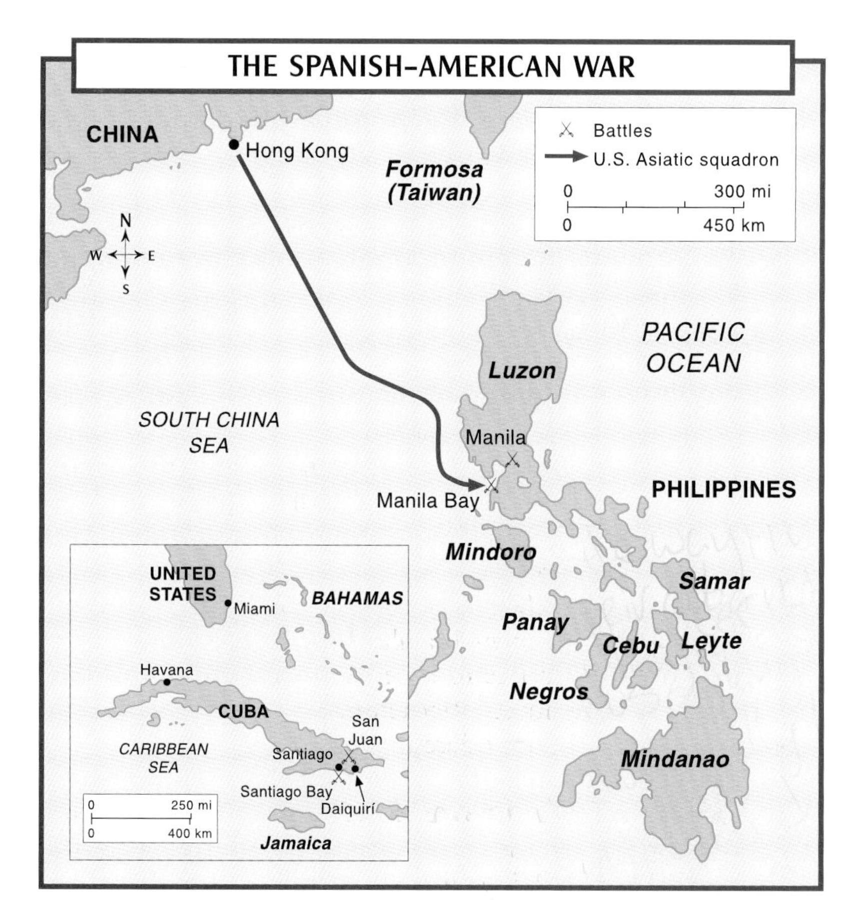

The 1898 war between Spain and the United States had two main theaters of operations, the Philippines (main map) and Cuba (inset). U.S. forces were victorious in both war theaters.

General William Shafter was chosen to lead the U.S. invasion force. He was given command of 17,000 troops, and these began landing at Daiquirí to the east of Santiago on June 22. The landings took three days, but went smoothly. The Spanish garrison on Cuba did not interfere. Most of the U.S. troops were regulars but there were units of volunteers present, including the First Volunteer Cavalry, better known as the "Rough Riders." Shafter had to cope with many problems—his cavalry units were short of horses, some vital equipment was lacking, and his men fell prey to various tropical diseases.

Santiago surrenders

Shafter moved his forces to besiege Santiago. The Spanish had 35,000 troops in the area but only 13,000 were holding the city. Before Shafter could close on Santiago, however, he had to deal with the 1,200 Spanish holding San Juan, a ridge running across

U.S. troops come under heavy Spanish fire as they assault San Juan Hill outside Santiago, on July 1, 1898. Some 1,500 Americans were killed or wounded before the hill was captured.

the road to Santiago. One part of the ridge, Kettle Hill, was the key position. Shafter decided to attack Kettle Hill and a second Spanish position, El Caney, at the same time.

The attacks began on July 1 and both Kettle Hill and El Caney were captured. The charge of the "Rough Riders," who were led by future U.S. president Theodore Roosevelt, was singled out by the U.S. press for praise, although it was an African-American unit, the U.S. Tenth Cavalry led by Lieutenant John J. Pershing, that bore the brunt of the fighting for the hill. The Spanish suffered a little over 800 casualties in the attack; the U.S. forces 1,500. The capture of the ridge, accomplished by the evening, decided the fate of Santiago.

A running fight

If the fall of San Juan was a major blow to the Spanish, what followed was a disaster of equal size. On July 3 Cervera decided to lead his squadron out of Santiago Bay to prevent it from falling into U.S. hands. The U.S. Atlantic Fleet was waiting for him. Sampson was away from the fleet so command rested with Commodore Winfield Schley. Seeing the advance of the Spanish fleet, Schley ordered his own warships to steam straight for the enemy. In the running fight that followed, the U.S. warships fired 8,000 shells at the Spanish. Less than 200 hit their targets, but they were sufficient to destroy the Spanish fleet. All six of the Spanish vessels were forced to run aground by the heavy fire directed against them by the U.S. warships. U.S. casualties in the battle were light—one man killed and one wounded. The Spanish had about 500 sailors killed and wounded, and nearly 1,800 men were taken prisoner.

The Battle of Santiago Bay effectively ended the war. Santiago surrendered on July 17, and the West Indian island of Puerto Rico followed suit on August 25. The end of the war was confirmed by the Treaty of Paris, which was signed on December 10. Spain was forced to give Cuba its independence and cede the Philippines, Puerto Rico, and the Pacific island of Guam to the United States for $20 million. The United States had gained great prestige from the war and become a major power in the Pacific. Spain, one of the first colonial powers, had lost its last colonies in the Pacific and Latin America.

The wreck of the Spanish warship Almirante Oquendo *lies beached off the southern coast of Cuba after the U.S. Navy's victory in the Battle of Santiago Bay on July 3, 1898.*

The Anglo–Boer Wars

By the second half of the 19th century the British were pushing northward into the interior of southern Africa from their Cape Colony and Natal. Much of the lands next to the Cape Colony were occupied by farmers of Dutch descent, the Boers, who had founded two states, the Orange Free State and the Transvaal. The discovery of valuable gold and diamond deposits in these areas encouraged the British to try to take them over. This greed led to the British fighting the Boers in two wars.

The First Anglo–Boer War was prompted by Britain taking control of the Transvaal in April 1877, although fighting did not begin until the Boers retaliated by forming their own republic at the end of December 1880. The Boers began by attacking British troops and laying siege to British garrisons, but the main action took place around Laing's Nek, a pass through the Drakensberg Mountains that separated the Boers' republic from British-held areas. Few troops were involved but the Boers, thanks to their superior marksmanship, usually bested the British.

Boer farmers take up position at Laing's Nek in 1881, during the first of their two wars with the British.

The Boers did not have a trained army, but all Boer males were expected to fight. Most were excellent shots and skilled horsemen. They were highly mobile, and usually fought from mountaintops, or in trenches. The British, who attacked in close ranks and were still wearing red uniforms, made easy targets. The first British attack on Laing's Nek (January 28, 1881) was defeated, but the fighting continued.

On February 27, the British tried to get behind the Boer positions on Laing's Nek by climbing a flat-topped mountain known as Majuba during the night. From the mountaintop, the British could fire down on the Boer positions in the distance. However, some Boers were able to climb up the steep, rock-covered slopes of Majuba without being spotted by the British on the mountain's summit.

Once at the top, the Boers were able to take some unoccupied and slightly higher ground that allowed them to fire down on the nearby British. The British, some 500 men, were virtually destroyed—nearly 100 were killed, over 130 were wounded, and about 60 captured. The defeat at Majuba led to a peace treaty, which recognized the right of the Boer Republic to exist. It was signed on April 5, 1881.

BOER COMMANDOS

Apart from a few paramilitary police and trained artillery units, the Boers, who were mostly farmers, did not have a regular army. However, men of military age were expected to report for service in wartime. They were required to bring with them a horse, rifle, ammunition, and food. They did not have uniforms, but wore their ordinary, everyday clothes.

The Boers fought in local units, usually containing men from the farmsteads around a particular town. These units were known as commandos and each had a strength of a few hundred men on average. The Boers did not have trained officers as such, their military leaders were elected by their men, and some were naturally talented commanders.

Despite being heavily outnumbered by the British, the Boers had several advantages. They were all mounted and could move at great speed. Most were excellent marksmen who had sharpened their skills hunting. They also knew the geography of southern Africa much more clearly than the British. Finally, if a battle was going badly, they rode away to fight another day. This made it difficult for the British to destroy them in a single battle.

The Second Anglo–Boer War

Defeat in the First Anglo–Boer War did not end Britain's determination to take over the Boer lands. In December 1895, Dr. L. Starr Jameson, a close friend of Cecil Rhodes, head of the Cape Colony, led 500 men into the Transvaal hoping to provoke a rebellion against the Boers. The expedition was a complete failure, but it further soured relations between the Boers and the British. In 1899, the president of the Boer Republic, Paul Kruger, issued a demand that British forces massing in Natal for a possible invasion should be withdrawn. The British refused, the Orange Free State sided with the Boer Republic, and war broke out in October.

On paper at least, the Boers appeared to stand little chance. The Boers had no army to speak of and could put fewer than 40,000 men into the field at any one time, although this total was greater than the troops the British had in southern Africa at the outbreak of war. The British had a highly professional modern army and could also draw on the military resources of the British Empire. The Boers, however, intended to strike quickly before the British could rush reinforcements to southern Africa.

British cavalrymen move out of their camp to fight against the Boers. The British were particularly short of cavalry units in the early part of the Second Anglo-Boer War (1899–1902), which gave the mounted Boers the freedom to move around at will.

British disasters

The Boers, all mounted, attacked three British garrisons. Boers from the Transvaal under Piet Cronjé laid siege to Mafeking, while Boers from the Orange Free State surrounded Kimberley. The largest Boer force, commanded by Petrus Joubert, won four battles in October—Laing's Nek, Talana Hill, Elandslaagte, and Nicholson's Nek—and besieged Ladysmith, the third garrison.

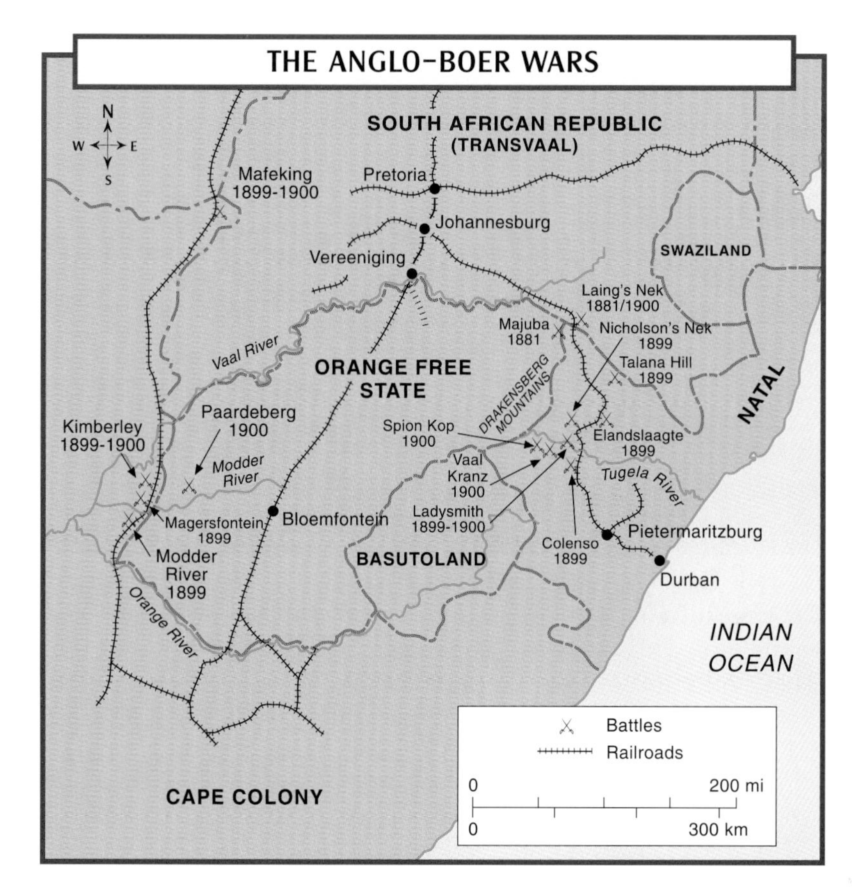

The British fought two wars against the Boers in the last decades of the 19th century. In both cases they inflicted serious defeats on the British.

The British response to these attacks was far too ambitious. Despite having few soldiers immediately available, the British decided to relieve all three towns at the same time. One force, 10,000 men commanded by General Lord Paul Methuen, struck out for Kimberley. Some 7,000 Boers commanded jointly by Cronjé and Jacobus de la Rey rushed south to oppose Methuen.

The Boer commanders' plan was to occupy hilltop positions in front of the British and then cut them down with rifle fire as they attacked across the open plains beneath them. Methuen launched a frontal attack on the Boers, who were in position along the banks of the Modder River, on November 28. It was a failure. On December 10–11, Methuen threw his forces against the Boers at Magersfontein and his forces lost 1,000 men killed, wounded, or missing (deserted or taken prisoner).

At the same time as Methuen's forces were being cut up by the Boers, General Sir Redvers Buller was leading over 20,000 British troops to Ladysmith. He, too, met with disaster. At the Battle of

A view of the Battle of Paardeberg in February 1900 showing British artillery bombarding the Boer camp. The camp was hemmed in on three sides by a river, which prevented the Boers from escaping. The balloon was used to identify targets for the British artillery overlooking the Boer camp.

Colenso the Boers did not dig in on top of hills, but at their foot. While the British advanced across the open plain before the Boers, their artillery was firing at the non-existent Boer positions on the hilltops. The British suffered nearly 1,000 casualties while the Boer reported only 50 men killed or wounded. British newspapers christened the series of December defeats "Black Week," and the government had no option other than to replace Buller.

Overall command of the British was given to Field Marshal Sir Frederick Roberts in January 1900. Roberts saw that the Boers' chief asset was their mobility. He decided to increase the proportion of cavalry in his own forces. He drew on the horsemanship of Canadian and Australian soldiers. While this buildup was taking place, there was more bad news. Buller was still trying to reach Ladysmith in early 1900 but his forces were halted by the Boers at Spion Kop (January 23) and Vaal Kranz (February 5).

Fighting to the bitter end

Roberts, however, was by now ready. He set out for Kimberley in late January. In the middle of February part of his army advanced against the Boers blocking the most direct route to Kimberley, while he led the remainder around the flank of the Boers, forcing them to retreat. Kimberley was relieved on the 15th.

On the 18th, the British caught up with the Boers at Paardeberg. The British settled down to a siege of the Boer camp and forced Cronjé with some 5,000 Boers to surrender on the 27th. A day later, at the third attempt, Buller finally reached the garrison at Ladysmith. Mafeking, the last British garrison under siege by the Boers, was reached on May 17–18. Mafeking had been under Boer siege for some seven months.

The British now struck into the Boer heartland, capturing the key cities of Johannesburg (May 31) and Pretoria (June 5). The war seemed over. However, some Boers, known as "bitter-enders" because they wanted to fight on, continued the war. They waged a guerrilla campaign, launching raids on British outposts, ambushing small columns of troops, and destroying railroad lines. Their war was to last for another 18 months.

The British responded in three ways. First, they used fast-moving columns of cavalry and mounted infantry to chase the Boers. Second, they built a huge network of small forts connected by barbed wire, to deny the Boers their freedom of movement. Third, most controversially of all, they began destroying Boer farms and herding women and children into camps, thereby denying the Boer fighters the local support they needed to survive. However, the bitter-enders carried on fighting until the end of May 1902, when a peace treaty was signed at Vereeniging.

CONCENTRATION CAMPS

A controversial aspect of Britain's strategy to defeat the Boers was the use of concentration camps. These were filled with old men, women, and children who had aided the Boer troops, or were suspected of doing so. Before they were herded into the camps, the farms of these civilians were destroyed.

The camps were unhealthy places. Sanitation was at best inadequate, water was often in short supply or tainted, and what food was provided was of poor quality. Little medical aid was available. Some 120,000 people were kept in the camps and probably 20,000 of them died through malnutrition or illness.

A few Britons attempted to improve conditions in the concentration camps. The most important of these was a woman, Emily Hobhouse. She visited many of the camps and began a public campaign in Britain to improve conditions. An all-woman body, the Fawcett Commission, went to southern Africa and published a damning report in December 1901. The conditions in the camps improved and their death rates declined dramatically.

British soldiers separate mothers and children in a concentration camp.

The Boxer Rebellion

By the late 1800s Chinese resentment at the involvement of various foreign powers in their affairs let to a short but bloody war. A Chinese secret society, the Boxers, believed that countries such as Britain, Germany, Japan, and Russia were exploiting China. The Boxers began to attack targets, particularly Chinese Christian converts and foreigners. When the war against the foreigners began in mid-1900, the Boxers laid siege to a small area of Beijing, the Chinese capital, where the various foreign embassies were located.

The Chinese dowager empress, Tzu Hsi, told the foreign powers that she was trying to deal with the Boxers, but members of her inner circle at the Manchu dynasty court were actually helping them in secret. Senior court officials were convinced that the foreigners, through the trading rights they had forced on the Chinese, were making fortunes out of China.

The rebellion begins

Boxer attacks intensified in the summer of 1900. On June 9, part of Beijing's racecourse was burned down, and the Japanese ambassador was murdered two days later. The various foreign legation staffs (representatives of overseas governments) began to fortify their embassies. On the 19th, the German ambassador was assassinated. In the late afternoon of the following day the legations came under Chinese fire for the first time.

A relief force led by a British admiral, Sir Edward Seymour, was already advancing to the aid of the legations by June 20.

Boxers attack a Chinese convert to Christianity and destroy Beijing's telegraph system at the outbreak of their uprising in 1900.

THE BOXERS

The Boxers, or Righteous Harmonious Fists, were members of a Chinese secret society probably founded in the 1700s. By the late 1800s, they had many sympathizers in the Chinese court who, like the Boxers, resented the presence of foreigners on Chinese soil. There were two main areas of conflict. First, foreign missionaries were having increasing success in converting Chinese to Christianity. Second, several foreign powers had colonized key ports on the Chinese coast from which they monopolized China's trade for their own benefit.

In the last decades of the century, against a background of flood and famine, the Boxers began to whip up unrest among the ordinary Chinese, blaming their poverty on the foreigners' ruthless exploitation of China's resources.

The Boxers began to attack foreigners and Chinese Christians. They used traditional weapons, such as swords, spears, and shields, and believed that they could not be harmed by modern bullets. The Chinese authorities did nothing to protect foreigners and on January 11, 1900, stated that the Boxers were part of legitimate Chinese society and not the criminals they were considered to be by the representatives of the foreign powers.

The Boxers were not equipped to take on the modern forces fielded by the foreign powers and suffered heavy losses during several battles. When the war ended in Chinese defeat in 1901, part of the peace settlement imposed on the Chinese demanded the disbandment of the Boxers and made attacks on foreigners a crime punishable by death.

Seymour had 2,000 troops, chiefly British, German, and Russian, and had left his base near the Taku Forts, built by the Chinese to protect the mouth of the Pei-Ho River, on June 9. His first objective was Tientsin, from where he hoped to travel to Beijing by train. Beijing was some 100 miles (160 km) to the north.

Seymour never reached Beijing. Boxer attacks, destroyed railroad tracks, and a lack of supplies halted his advance some 30 miles (48 km) from the capital, at the village of An Ting. The relief column, fighting off Boxer attacks supported by units of the Chinese regular army, fell back to Tientsin, reaching the outer suburbs of the city on the 26th. However, the unrest had spread to Tientsin and the Boxers were already in control of parts of the city. Seymour took up a position just outside Tientsin.

While Seymour was making his moves, other international forces were arriving off the mouth of the Pei-Ho. The various commanders decided to attack the Taku Forts and then advance

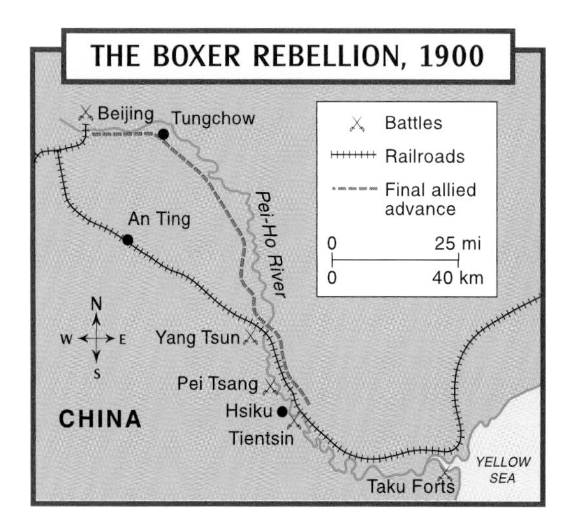

The route taken by the second foreign relief force that succeeded in reaching the besieged embassies in Beijing during the Boxer Rebellion.

to the aid of Tientsin. The dried-mud forts were captured on the 17th and part of the international force rushed to Tientsin. The European quarter of the city was under attack, but the Boxers and Chinese army soldiers were being held back in no small part due to the defenses prepared by a young engineer, Herbert Hoover, later president of the United States.

After some hard fighting and heavy losses on both sides, Tientsin was finally cleared of Chinese forces by July 23. Plans were laid to move on Beijing. The international force consisted of 20,000 troops, including some 10,000 Japanese, 4,000 Russians, 3,000 British, 2,000 Americans, 800 French, 200 Germans, 58 Austrians, and 53 Italians.

The embassies under attack

The advance north was extremely difficult for the relief force. High temperatures, bad roads, and Boxer attacks made for slow progress. On August 5, there was a battle at Pei Tsang, less than 10 miles (16 km) north of Tientsin, and the following day the international forces fought a second battle a few miles north of Pei Tsang at Yang Tsun. By the 12th, the city of Tungchow, 12 miles (18 km) from Beijing had been occupied and it was decided to storm the Chinese capital two days later. However, no one was sure that the embassies had not fallen to the Chinese.

The embassies, defended by about 400 regular troops and various armed civilians, were holding out, but were suffering greatly from shortages of food, bad water, and Boxer attacks supported by regular Chinese forces. On June 23, the Boxers had set fire to market buildings hoping that the flames would spread to the embassies, and on July 13 they had exploded a mine under the French embassy. Sniper and artillery fire added to the discomfort of the embassies' defenders and the Europeans and Chinese Christians sheltering behind their walls and barricades.

The final battle to relieve the embassies began on August 14. Although the various foreign forces outside Beijing had been given precise timetables and objectives, the assault degenerated into a race. Each foreign force wanted the glory of reaching the embassy quarter first. The race was barely won by the British, but

U.S. artillery fires on the walls of Beijing on August 14, 1900, during the final battle to relieve the various embassies besieged by the Boxers.

the scramble to reach the embassies clearly showed that the international alliance was falling apart as the various nationalities put their own interests first.

Ending the fighting

Over the following days the rest of Beijing was captured, and the allies then spread out from the Chinese capital to crush the remaining pockets of Boxer resistance. In late October the operations were brought to an end. The allies behaved particularly badly during the final weeks, looting and destroying many works of Chinese art. They also imposed a harsh peace treaty on the Chinese, including heavy fines. The Boxer society was disbanded as part of the final settlement agreed to on September 12, 1901.

The impact of the Boxer Rebellion eventually led to the downfall of the Manchu dynasty in 1912. It was the first time that the world's leading powers had acted together for their mutual benefit. However, the solidarity they showed in dealing with the Boxer Rebellion was not destined to last. Within five years of the beginning of the rebellion, the Japanese and Russians would be at war, and within 14 years virtually all of those who had contributed troops to the relief of the embassies in Beijing would be fighting each other in World War I.

THE RUSSO–JAPANESE WAR

By the beginning of the 20th century Japan held an increasingly dominant position in the northern Pacific. However, Japan wanted to expand its political, economic, and military influence in the region beyond the Korean peninsula, which it controlled. Any such moves would anger Russia, also an important power in the area. Japan decided to go to war in 1904. It had several objectives—capture Port Arthur, the chief Russian naval base, destroy the Russian Far East fleet, and smash the Russian army in Manchuria.

The Japanese did not issue a formal declaration of war but launched a surprise attack with torpedo boats on the Russian fleet as it lay at anchor in Port Arthur's harbor on February 8, 1904. Considerable damage was caused and the Japanese then settled down to blockade the port. War was declared two days later. The Japanese still held the initiative and followed up the Port Arthur attack by landing at Chemulpo on the Korean peninsula on February 17, 1904. The Japanese advanced north to the Yalu River, Korea's border with Russian-controlled Manchuria.

The blockade of Port Arthur continued but the arrival of a dynamic Russian naval officer, Admiral Stepan Makarov, threatened to break the stalemate. Makarov launched a series of attacks, taking care to avoid battle with the main Japanese fleet under

Japanese warships open fire on the Russian artillery protecting Port Arthur, February 8, 1904. At the same time several torpedo boats attacked Russian warships at anchor in the port's harbor. The surprise attack took place before war had been declared.

The war between Russia and Japan saw the Japanese almost wholly successful on land and at sea. The Russians lost and were forced to surrender much of their influence in the region to the victorious Japanese.

Vice-Admiral Heihachiro Togo. These Russian attacks were successful but Makarov's flagship, the *Petropavlovsk*, hit a mine on April 13. The entire crew and Makarov were killed. It was a major blow to the Russians. Morale in Port Arthur fell with the loss of Makarov, and the Japanese blockade stayed in place.

The Russians, aware of the threat to Port Arthur and of the Japanese landings in Korea, tried to gather their forces. The Russian commander, General Alexei Kuropatkin, knew he was outnumbered and wanted to build up his forces. However, Kuropatkin's superior, Admiral Evegeni Alekseev, disagreed. He demanded action. The folly of this plan became clear in late April. The Japanese overwhelmed a small Russian force at the Battle of the Yalu River.

The Japanese kept up the pressure on the Russians in the first part of May. Their Second Army landed only 40 miles (64 km) northeast of Port Arthur and began to advance on the base. The Japanese Fourth Army then landed a little to the west of the Yalu.

The Japanese Second Army had to defeat the Russians holding Hashan Hill, a vital part of Port Arthur's outer defenses. The battle on May 25 saw 3,000 Russians take on 30,000 Japanese. The Japanese launched wave after wave of frontal assaults and their casualties were heavy (4,500 men), but the hill was captured. The capture of the hill also allowed the Japanese to use a nearby port as a naval base. The Japanese Third Army began building up there. The siege of Port Arthur began in June 1904.

June saw the Russians attempt to prevent the Japanese buildup by attacking their warships. On the 15th, the Japanese lost two warships to mines, although a halfhearted Russian attack was thrown back on the 23rd. An attack by Russian land forces was also blocked on the 26th.

This operation marked the end of Russian attempts to break the siege. The Russian ships at Port Arthur were ordered to flee, but the attempt ended in disaster at the Battle of the Yellow River on August 10. Admiral Vilgelm Vitgeft had 19 warships. Caught by the Japanese under Togo, Vitgeft's ships were pounded for 90 minutes. One was sunk and the surviving ships, many damaged, sought sanctuary in neutral ports or scurried back to Port Arthur.

Japanese officers survey the harbor of Port Arthur shortly after the Russian base had fallen into Japanese hands in January 1905. Note the sunken Russian warship lying on the harbor bed.

Port Arthur under fire

The Japanese began their land attacks against Port Arthur in early August but suffered very heavy casualties. The Japanese needed siege artillery. The heavy artillery arrived at the beginning of October and bombarded the Russian positions around the clock. Despite the barrage a Japanese attack at the end of the month also failed, as did another in November. Casualties were heavy.

The fighting centered on a Russian position known as 203 Meter Hill. If it fell, Port Arthur and the remains of the Russian fleet would be at the mercy of the Japanese. The final battle for the hill lasted from November 27 to December 5. The Japanese took the position. Port Arthur could not survive and the Russians surrendered on January 2, 1905.

While the siege was progressing, Russian troops tried to smash the Japanese forces in Manchuria. Their early attacks in

1904 were repulsed at two battles in the second half of July and the Russians were forced to fall back. The Japanese followed up and fought an inconclusive battle at Liaoyang, which ended on September 3. The Russians began to fall back on Mukden, their center of operations. Two battles in the spring of 1905 were also inconclusive, but the next battle decided the land war.

The Battle of Mukden was huge. Both sides had more than 300,000 troops spread out and entrenched over a distance of 40 miles (64 km). The fighting began on February 21 and lasted until March 10. The Japanese won a narrow victory.

Catastrophe at Tsushima

The fighting on land was over, but the Russians suffered one more disaster at sea. The Battle of Tsushima, fought on May 27–28 between the Russian Baltic fleet led by Admiral Zinovi Rozhdestvenski and a Japanese fleet commanded by Togo, remains one of the most decisive naval engagements of all time. The Russian Baltic fleet had sailed half way around the world only to be annihilated. Togo's warships outgunned and outmaneuvered the Russians. The Russian fleet of eight battleships, eight cruisers, nine destroyers, and a number of smaller craft was smashed by long-range gunnery and attacked by torpedoes in the two-day battle. Only one cruiser and five destroyers escaped.

Japanese admiral Heihachiro Togo (left), commander of the Japanese fleet at the Battle of Tsushima in 1905, pictured on the deck of his flagship during a storm.

After such catastrophic defeats on land and sea, the Russians had to agree to the terms of the Treaty of Portsmouth, which was arranged by U.S. President Theodore Roosevelt, on September 6, 1905. Japan took over Port Arthur and other territories, the Russians had to withdraw their troops from Manchuria, and Korea was recognized as being firmly controlled by Japan. The Japanese victory was, however, more significant than these territorial exchanges suggest. At a stroke the Japanese had become the dominant power in the northern Pacific. In less than 40 years Japan would attempt to become the dominant power across the whole of the Pacific.

War in The Balkans

At the dawn of the 20th century the Ottoman Turkish Empire was in decline and several of the countries it controlled in the Balkans were gripped by nationalist feelings and sought their independence. In 1911 Italy declared war on Turkey in an effort to carve out its own empire in North Africa by occupying what is now Libya, then a Turkish possession. Taking advantage of this, three Balkan states—Bulgaria, Greece, and Serbia—formed the Balkan League and declared war on Turkey in late 1912.

Turkey was overstretched. The Balkan League, supported by Montenegro, could muster about 350,000 troops, while the Turks had fewer than 250,000 men available. The League launched several attacks against the Turks in October 1912. The Greeks under Crown Prince Constantine attacked and, despite suffering some setbacks, defeated the Turks on November 5. The Serbians also defeated the Turks at Kumanovo on October 24, and forced them to retreat to Monastir.

Aided by a searchlight used to spot targets, Bulgarian artillerymen open fire on the Turkish city of Adrianople during February 1913.

The Battle of Monastir on November 5 was a hard-fought contest with both the Serbians and Turks showing great bravery. Gradually, however, the Serbs launched a frontal attack on the Turks and broke through. Threatened by a Greek army advancing from the south, the Turks retreated. The Greeks captured the fortress of Salonika four days later and placed a number of other Turkish garrisons, including Scutari, under siege.

Balkan League victories

The Turks were faring no better elsewhere. Three Bulgarian armies advanced on a broad front and defeated the Turks at Seliolu and Kirk Kilissa at the end of October. The Turks fell back to hold a 35-mile (56-km) long defensive line between Lülé Burgas and Bunar Hisar. Two of the Bulgarian armies chased eastward after the Turks, while the third placed the city of Adrianople to the west under siege.

The Bulgarian attacks on the Turkish defensive line at Lülé Burgas on October 28–29 were successful and the Turks had to pull back even farther. They took up a position along the Chatalja Line, their last defensive barrier before Constantinople. The Bulgarians tried to smash through this Chatalja Line during

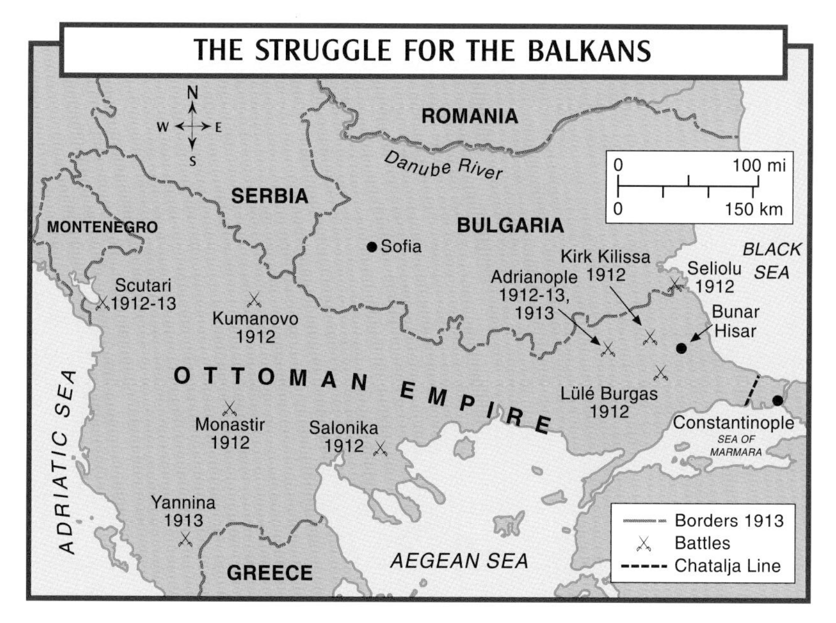

Two wars were fought in the Balkans between 1912 and 1913. In the first, three Balkan states formed an alliance to fight Turkey. The second involved the original three Balkan states, which went to war against each other.

November and December, but it proved too strong for them. The Turkish capital was safe. Peace talks began in the middle of December and an armistice brought the war to a temporary halt.

However, the Turkish government was overthrown in January 1913 and replaced by an extreme nationalist group known as the "Young Turks." They wanted the war to continue. Despite their efforts, the Turkish armies suffered further defeats in 1913. The Turkish cities of Yannina (March 3), Adrianople (March 26), and Scutari (April 22) all fell to the Balkan League. The Turks sued for peace. The Treaty of London saw Turkey lose virtually all of its possessions in the Balkans.

An uneasy peace

The Balkan League did not survive victory in the First Balkan War. National rivalries soon tore the alliance apart. In May 1913, Bulgaria declared war on both Serbia and Greece, hoping to gain more territory. Bulgaria severely underestimated the strength of its enemies, and its invasions were repulsed. Romania sided with Serbia and Greece, and the Turks took the opportunity to recapture Adrianople. The Treaty of Bucharest ended this Second Balkan War on August 10. Bulgaria was forced to give up most of the territory it had gained during the war against Turkey. The peace in the Balkans did not last long. Within a year an assassination in the Balkans sparked the outbreak of World War I.

GLOSSARY

battleship The most powerful warship afloat in the 19th century. Typically, a battleship was either sail- or steam-driven, was made of wood covered in metal plates or wholly of metal plates. Main armament usually consisted of two or three guns placed in rotating armored turrets at either end of the vessel.

breechloader A rifle or artillery piece that is loaded at the rear of the barrel rather than down the muzzle. Troops using breechloaders were able to load and fire much more quickly than those using muzzle-loading weapons.

colony A country taken over and ruled by another. Many of the wars in the second half of the 19th century involved European powers establishing and defending their colonies around the world.

guerrilla warfare A type of warfare fought by outnumbered forces against large armies. The guerrillas try to avoid battles against larger enemy forces. Instead they launch hit-and-run raids on the enemy, carry out ambushes, and strike at the enemy's weak points. Guerrillas aim to undermine an enemy's will to win rather than inflict a major single defeat on their opponent.

machine gun A quick-firing weapon able to fire hundreds of rounds of ammunition every minute. They were developed in the second half of the 19th century and ensured that infantry units could no longer make charges without suffering massive casualties. Early machine guns were mounted on wheeled carriages; later ones on a folding tripod.

mobilization The process by which trained soldiers prepare for war. Reservists return to their unit depots, are given weapons, equipment and other vital supplies, and are then transported to the battlefront.

reservists The name given to soldiers who have undergone military training but have returned to civilian life. They undergo a few days or weeks of training each year and have to return to their units during times of war.

staff A group of highly trained officers, usually specialists in a particular aspect of warfare, such as supply, troop movement, and logistics. These officers support field commanders and senior generals by planning wars and campaigns, and try to ensure that the fighting follows the plans that they, the staff, have devised.

BIBLIOGRAPHY

Note: *An asterisk (*) denotes a Young Adult title.*

*Brownstone, David and Franck, Irene. *Timelines of Warfare From 100,000 B.C. to the Present.* Little, Brown and Company, 1994.

Chandler, David G. *Atlas of Military Strategy—The Art, Theory, and Practice of War, 1618–1878.* Sterling Publishing Co., Inc., 1998.

Greene, Jack, and Massignani, Alessandro. *Ironclad at War—The Origin and Development of the Armored Warship, 1854–1891.* Combined Publishing, 1998.

Howard, Michael. *The Franco–Prussian War: The German Invasion of France, 1870–71.* Routledge, 1998.

*Keegan, John, and Wheatcroft, Andrew. *Who's Who in Military History: 1453 to the Present Day.* Routledge, 1998.

*Knight, Ian. *Great Zulu Battles, 1838–1906.* Sterling Publishing Co., Inc., 1998.

Nofi, Albert A. *Spanish–American War, 1898.* Combined Publishing, 1996.

Spiers, Edward M. *Sudan: The Reconquest Reconsidered.* Frank Cass Publishers, 1998.

INDEX

ACKNOWLEDGMENTS

Cover (main picture) Peter Newark's Western Americana, (inset) Peter Newark's Military Pictures; page 1 Peter Newark's Military Pictures; page 5 AKG Photo, London; page 6 AKG Photo, London; page 8 AKG Photo, London; page 10 Peter Newark's Military Pictures; page 12 Peter Newark's Military Pictures; page 13 Peter Newark's Historical Pictures; page 14 Robert Hunt Library; page 15 Peter Newark's Military Pictures; page 16 Peter Newark's Military Pictures; page 18 Peter Newark's Military Pictures; page 20 AKG Photo, London/Erich Lessing; page 21 AKG Photo, London; page 23 AKG Photo, London; page 25 AKG Photo, London; page 26 Peter Newark's Military Pictures; page 28 AKG Photo, London; page 30 Peter Newark's Military Pictures; page 31 Peter Newark's Military Pictures; page 32 AKG Photo, London; page 33 AKG Photo, London; page 34 AKG Photo, London; page 36 Peter Newark's Military Pictures; page 37 Peter Newark's Military Pictures; page 39 Robert Hunt Library; page 40 Peter Newark's Westerm Americana; page 42 Peter Newark's Western Americana; page 43 AKG Photo, London; page 45 AKG Photo, London; page 46 Peter Newark's Western Americana; page 48 Peter Newark's Military Pictures; page 51 AKG Photo, London; page 52 Peter Newark's Military Pictures; page 55 AKG Photo, London; page 56 Popperfoto; page 58 Peter Newark's Military Pictures; page 60 Peter Newark's Military Pictures; page 61 Brown Partworks; page 62 Peter Newark's Military Pictures; page 64 AKG Photo, London; page 66 AKG Photo, London; page 67 AKG Photo, London; page 68 AKG Photo, London; page 71 Brown Partworks; page 72 AKG Photo, London; page 74 AKG Photo, London; page 75 Peter Newark's Military Pictures; page 76 AKG Photo, London.